What Every Engineer Should Know About Inventing

WHAT EVERY ENGINEER SHOULD KNOW
A Series

Editor

William H. Middendorf

Department of Electrical and Computer Engineering
University of Cincinnati
Cincinnati, Ohio

Other volumes in preparation

AN IMPORTANT MESSAGE TO READERS. . .

A Marcel Dekker, Inc. Facsimile Edition contains the exact contents of an original hard cover MDI published work but in a new soft sturdy cover.

Reprinting scholarly works in an economical format assures readers that important information they need remains accessible. Facsimile Editions provide a viable alternative to books that could go "out of print." Utilizing a contemporary printing process for Facsimile Editions, scientific and technical books are reproduced in limited quantities to meet demand.

Marcel Dekker, Inc. is pleased to offer this specialized service to its readers in the academic and scientific communities.

What Every Engineer Should Know About Inventing

William H. Middendorf

Department of Electrical Engineering
and Computer Science
University of Cincinnati
Cincinnati, Ohio

CRC Press
Taylor & Francis Group
Boca Raton London New York

CRC Press is an imprint of the
Taylor & Francis Group, an informa business

Library of Congress Cataloging in Publication Data

Middendorf, William H.
  What every engineer should know about inventing.

  (What every engineer should know ; v. 7)
  Bibliography: p. 142
  Includes index.
  1. Inventions. 2. Creative ability in technology.
I. Title. II. Series.
T212.M52   608   81-12469
ISBN 0-8247-7497-3   AACR2

MARCEL DEKKER, INC.
270 Madison Avenue, New York, New York 10016

Current printing (last digit):
10 9 8 7 6 5 4 3 2

Foreword

When I began to think about what to say about a book on invention, I thought it would be proper to point out that the state of invention and particularly the innovation that follows invention is in an unsatisfactory condition in the United States. But the more I thought of it, the more I realized that our illness is a disease of all large organizations, of management of technical enterprises by nontechnical people, of reliance on rules and regulations rather than human ingenuity, and that it is not limited to the United States. The problems that befall an original thinker under our system are not different or may actually be easier than the problems that befall an inventor in a socialistic country, the largest of large organizations.

From conversations with scientists and Patent Office officials of Russia, I learned that as organizations get larger and larger the procedures, promoters, and managers become similar in the capitalistic and socialistic countries alike. The inventor who is basically an individual suffers from the fact that he becomes a cog in a big machine and a cog whose work and output is not understood and very often not appreciated by the management in the many levels above him.

It is a fact of human intelligence that the art of invention, like the art of anything else, is based on new, unexpected, and unobvious combinations of old ideas. Such new and nonobvious combinations do not, of course, appear except in a single brain. They can later be analyzed and dissected by others and the others, inspired by the new thought, may then produce different and new combinations. Inventing, like composing music, writing poetry, and creating a new painting, remains the act of a single set of neurons. It is, of course, true that sometimes an invention may have more than one part and more than one human being may contribute to it, but the really basic, simple, elegant idea is born in one person's brain; this book is devoted to the procedures, the incentives, and some of the rewards of this activity.

Just reading this book will not make everyone an inventor, but those who have the talent and the desire will learn much from it. The examples and procedures described certainly help the reader to understand the process and to be encouraged by the understanding.

This book is addressed primarily to the young inventor and this is as it should be. Young people are relatively free of too much tradition, too many ingrained habits, and too many activities that detract from their ability to do original work. Some of us are lucky and we continue to invent in spite of the fact that we become managers of enterprises, parents of children, and have other distractions from the work of which we are capable. While there is much evidence that we do slow up, at least mentally, as we get older, there is considerable evidence that we make up for this by knowing more science and being more realistic about the value of our inventions so that we can better separate the wheat from the chaff. It must always be remembered that one cannot think only of the successful and good ideas, because invention, like other acts of creativity, is basically a random process of our brain. A great deal of chaff must be produced, and it is the function of our talent, our training, our realism, and our intellectual honesty to separate the wheat from the chaff and concentrate only on those ideas that give promise of intellectual and physical rewards.

This book is full of examples of brilliant work. Some of the inventions described are startling in their simplicity and these are the inventions that have true beauty. What can one say to the aspiring inventor? All one can say to those who read this book is "go thou and do likewise."

Jacob Rabinow
National Bureau of Standards
Washington, D. C.

Preface

This book has evolved from my interest in invention during most of my professional career. In the early 1950s my research and consulting work resulted in several patents. About that same time a plethora of articles on creativity began to appear in publications reflecting the results of research conducted during the previous decade. This information and my experiences were incorporated in a senior design course that has expanded as my knowledge of the subject has matured.

The material in this book was selected to fulfill the promise of the title. It is truly meant to be what every engineer should know about the subject. It is not meant to be all that can be written about it. The book is also meant to be truly effective in giving you, the reader, the information you need to develop into a person who seeks creative opportunities and responds with elegant inventions. Explicit advice is included on how to increase creativity by working at it. I have no magic formula. The difficult follow-up work which takes an invention to production is not neglected nor is the message that that phase may be as difficult as the invention.

Liberal use is made of examples to impress on your mind the lessons to be learned. As best as can be determined, these are true stories. They have been accumulated from experience, from attending lectures on inventing, and from other written work or private correspondence. A chapter of short case studies is included to introduce you to a variety of inventors and to learn how they describe the genesis of their inventions.

The book was purposely written to be easily understood by all engineers regardless of the area of their expertise. Although the examples may involve technical details, these are never important or complex enough to obscure the lesson to be learned from the example. Also, students at the junior and senior levels will have no difficulty in learning what this book is meant to teach. Since courses specifically designed to promote invention are unusual in engineering college programs, I hope that teachers who agree with my conviction that invention is necessary for continued technological progress will find this book suitable to assign as outside reading in design or project-related courses.

I wish to acknowledge the help of a few of the many people who have influenced the writing of this book; some very casually, others more directly. It was an offhand remark by John L. Baker, now Professor Emeritus of Mathematics, at a luncheon gathering that led me to Koestler's book, *The Act of Creation.* This book more than any other publication is woven into the thoughts expressed here. It was the Wadsworth Electric Manufacturing Company, Inc. which retained me for a time specifically to establish a patent position in their lines of manufacture. This gave me the opportunity to practice the art of inventing. It was the ability to pick up the telephone at any time and discuss invention or inventors with Mr. William G. Konold or Donald F. Frei, two of the coauthors of a companion book in this series (on patents), which provided accurate information that only patent attorneys with broad experience are likely to know.

A major advantage has been teaching the techniques of product development to students in the University of Cincinnati cooperative (work/ study) engineering program. The students' enthusiasm to learn about inventing and the valuable suggestions made by the classes of 1979 and 1980 when the manuscript was in final production provided the encouragement needed to bring this book to completion.

Finally, a word about gender. This book speaks to women as well as men. An attempt was made to use both genders for each pronoun but it was abandoned as being inconvenient for the reader. Be assured that when I say "engineer" I mean male or female and when I say "he" or "him" I mean to include women.

William H. Middendorf

Contents

About the Author

William H. Middendorf is Professor of Electrical Engineering at the University of Cincinnati. He received a Bachelor of Electrical Engineering degree from the University of Virginia, an M.S. from the University of Cincinnati, and a Ph.D. from Ohio State. He has 27 patents and has written two prize-winning papers on design. He is a member of the General Engineering Committee of the Low Voltage Distribution Section of NEMA and a member of three Underwriters Laboratories Industry Advisory Councils. Dr. Middendorf is a Fellow of IEEE and received the Herman Schneider award as Distinguished Engineer in 1978. He is listed in national and international biographical publications including *Who's Who in Engineering* (1978) and *Who's Who in America* (1980).

What Every Engineer Should Know About Inventing

1

The Climate for Invention

It would be convenient in starting this study to simply assume that every engineer has the degree of motivation to become a modern day Edison. Convenient, but not realistic. As a matter of fact, indications are that the creative output of the scientific-engineering community has fallen during the last quarter century. Many reasons have been given. Some persons believe the government has not responded to the changing needs which encourage invention. Others argue that the attitude of industry discourages inventors.

However, the reasons for the decrease in creative productivity is much more complicated than can be explained by a few simple statements. This subject will be discussed at some length in this chapter to, as it were, give an account of the national climate for invention.

DEFINITIONS

Before proceeding it is appropriate that the various words which express the main thrust of this book be defined. They are *creativity, invention, innovation,* and *patent.*

Creativity is the ability to produce novel ideas or things which are unexpected and show a high degree of skill and intelligence. It applies to contributions in any field of human activity. Invention, as used here, means the process of devising and producing by independent investigation, experimentation, and mental activity something which is useful and which was not previously known or existing. An invention involves such high order of mental activity that the inventor is usually acclaimed even if the invention is not a commercial success. Often inventions are put into use after they become public property.

Innovation, which may or may not include invention, is the complex process of introducing novel ideas into use or practice and includes entrepreneurship as an integral part. Innovation is usually considered noteworthy only if it is a commercial success. Thus society benefits from innovation, not from invention alone, and often there is a significant lapse of time from invention to innovation.

Finally, a utility patent is a legal document issued by the U.S. government to encourage technological progress by giving a limited monopoly to the inventor. It requires that the inventor prove to the satisfaction of an examiner that the device or process claimed to be an invention is novel, useful, and nonobvious to those skilled in the art. Not every invention is protected by a patent. Some inventions are better protected as trade secrets and others, such as methods of doing business and new developments in mathematics, are specifically excluded from patent protection.

PATENT STATISTICS

All kinds of creativity are important to human development. However, the thrust of this book is technological development. This should include trade secrets and other creative output that contributes to technology. Data are not available for this spectrum of creativity. As a consequence, the rate of issue of patents will be used as a measure of present day creative output with full realization that it is not a perfect metric. There simply is no better one. When a freelance inventor or an inventor's employer decides to protect an invention by applying for a patent, there is a certain evaluation made that the expenditure is worth it. This kind of close evaluation balanced against cost gives confidence that patents represent the highest quality creative output.

Although the number of engineers holding bachelor, master, or doctoral degrees has increased dramatically during the last quarter century, the number of U.S. patents issued to U.S. citizens has not kept pace [1]. Furthermore, as shown in Figure 1.1, a larger and larger share of those issued have gone to foreign inventors. Overall the portion of U.S. patents issued to foreign inventors is approaching 40%. Even more disturbing is the fact that the foreign share of U.S. patents ranges from 50 to 85% in those areas identified by the Office of Technology Assessment as "high technology" [2]. It is not unusual to see pages of the *Patent*

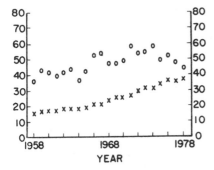

Figure 1.1 The dramatic increase of U.S. patents to foreign citizens compared to the slight increase in patents to U.S. citizens: (o), patents to U.S. citizens in thousands; (x), % total to foreign.

Gazette list a preponderance of foreign inventors as shown in Figures 1.2 and 1.3.

THE CHANGE IN EDUCATION

Engineering education very properly subjects itself periodically to self-evaluation. One of these periods of analysis occurred in the late 1940s. In response to the comparison during World War II of the ability of engineers and scientists to adjust to new challenges as they were displaced from civilian jobs to the design and manufacture of war material, it was decided that engineering education should have a greater scientific bent. Curricula were changed to include more courses in analysis. Laboratory experience was reduced. Design courses were all but eliminated in some schools, and undergraduate thesis or senior development projects were included in fewer engineering programs. About this same time such agencies as NASA and NSF increased support of university research. This was largely scientific vis-a-vis engineering and was used as the basis for expanding graduate programs. Faculty shifted attention to research and graduate study because the rewards were there.

This change in faculty was accelerated as new appointments were made to the candidates most likely to engage in sponsored research. With each new face the students were taught more analysis techniques, more mathematical rigor, more advance physics concepts, but less about how products are made, how specifications are met, and how products fail.

said shifter; and a numerical positioning controller capable of controlling driving motors of said tool feed control devices.

4,224,848

CROSS CUTTER FOR ROLLS OF MATERIALS

Dieter Beerenwinkel, Düsseldorf, Fed. Rep. of Germany, assignor to Jagenberg Werke Atkiengesellschaft, Dusseldorf, Fed. Rep. of Germany

Filed Sep. 28, 1978, Ser. No. 946,689

Claims priority, application Fed. Rep. of Germany, Oct. 21, 1977, 2747256

Int. Cl.² B26D 5/24

U.S. Cl. 83—76   3 Claims

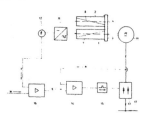

1. A cross cutter for a moving web, comprising coupled cutter rolls; a driving motor directly coupled to the cutter rolls for driving same; asymmetrical means driven by the driving motor in parallel with the cutter rolls for producing a control output corresponding to the actual number of revolutions per unit time of the motor; and means receptive of the control output for controlling the motor speed to synchronize the cutter rolls with the speed of the moving web.

4,224,849

DEVICE FOR DETECTION OF METAL IN A MOVED STRAND

Karlheinz Loser, Karlsruhe, Fed. Rep. of Germany, assignor to Weisert, Loser & Sohn GmbH & Co., Fed. Rep. of Germany

Filed Jan. 24, 1979, Ser. No. 6,197

Claims priority, application Fed. Rep. of Germany, Jan. 5, 1975, 2900280

Int. Cl.³ B65G 47/34

U.S. Cl. 83—80   9 Claims

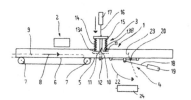

1. Device for detecting and removing metal from a moving strand of highly viscous material, such as gum or the like, comprising:

  a guide trough for the strand,

  an electronic metal detector for detecting metal in the strand as it moves along the guide trough,

  a drop knife disposed above the guide trough,

  control means for actuating said drop knife to sever said strand in response to detection of metal in the strand by the metal detector,

  and swing valve means opening into the floor of said guide trough for accommodating removal of the severed strand with metal therein.

4,224,850

APPARATUS FOR CUTTING A BLANK SHEET INTO STRIPS AND FOR STACKING THE STRIPS IN ADJOINING COMPARTMENTS SEPARATED BY PARTITIONS

Ismo V. J. Holmi, and Ismo I. Virtanen, both of Pori, Finland, assignors to Outokumpu Oy, Helsinki, Finland

Filed Feb. 23, 1979, Ser. No. 14,430

Claims priority, application Finland, Feb. 27, 1978, 780646

Int. Cl.³ B26D 7/18

U.S. Cl. 83—105   10 Claims

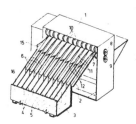

1. An apparatus for cutting a blank sheet into strips and for stacking these strips in adjoining compartments separated by partition walls, the apparatus comprising: a frame, at least two superimposed shafts mounted in the frame for rotation in opposite directions; slightly overlapping circular blades mounted in spaced relationship from each other in the axial direction on the shafts, in order to cut the blank sheet fed therebetween and to feed the cut strips in the direction of the feed downward; an inclined slide surface mounted in the frame for receiving the cut strips and having substantially on the same vertical longitudinal plane as the partition walls of the adjoining compartments, guide walls for receiving the cut strips and for guiding each of them into its respective compartment; lateral-transfer trays mounted on the inclined slide surface and extending, as seen in the direction of the feed, slightly forward and downward from the circular blades, being also inclined in the lateral direction and extending in the lateral direction from between two adjacent guide walls over one guide wall, at maximum as far as the vertical longitudinal plane running through a cutting point of one edge of the strip entering the lateral-transfer tray and at minimum through the vertical longitudinal plane which passes through a center point between the cutting points of the cut strip, in order to slant the cut strips and to cause that edge which is inclined lower to slide along the adjacent guide wall, so that the upper edge will fall between the guide walls before the strip falls into the compartment.

4,224,851

KNOCKOUT FOR PUNCH SCRAP

Hiroto Imai, Fukuyama, Japan, assignor to Mitsubishi Jukogyo Kabushiki Kaisha, Tokyo, Japan

Filed Jun. 26, 1978, Ser. No. 919,165

Int. Cl.³ B26D 7/00

U.S. Cl. 83—117   2 Claims

1. A scrap knockout for use with a rotary punching apparatus, comprising in combination:

  (a) a knife cylinder (35) with an outer peripheral surface;

  (b) a blade support (39) attached around said outer peripheral surface;

  (c) punch blades (37) held by and extending out of said blade support (39) with apertures (37') defined in said blades, said punch blades (37) defining a location for punched scrap;

  (d) an anvil cylinder (34) disposed for cooperation with said knife cylinder (35) having a periphery against which the blades (37) are forced so as to punch a work piece fed in between said knife cylinder and said anvil cylinder; and,

---

Figure 1.2   Page from *Patent Gazette* (U.S. Patent Office, September 30, 1980).

4,225,807
READOUT SCHEME OF A MATRIX TYPE THIN-FILM
EL DISPLAY PANEL
Masahiro Ise; Kenzo Inazaki; Katsuyuki Machino, all of Tenri, and Chuji Suzuki, Nara, all of Japan, assignors to Sharp Kabushiki Kaisha, Osaka, Japan
Filed Jul. 11, 1978, Ser. No. 923,646
Claims priority, application Japan, Jul. 13, 1977, 52-84352; Jul. 13, 1977, 52-84353; Jul. 27, 1977, 52-90660; Aug. 31, 1977, 52-105179; Sep. 6, 1977, 52-107423; Sep. 26, 1977, 52-116205
Int. Cl.² H05B 33/08
U.S. Cl. 315—169.3   5 Claims

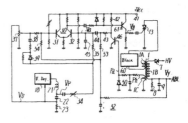

1. A display device having a plurality of picture elements comprising:
a thin-film EL display panel comprising an EL layer sandwiched by a pair of dielectric layers said EL display panel manifesting a hysteresis curve in the applied voltage vs brightness characteristics, and a matrix electrode array sandwiching said pair of dielectric layers for matrix driving said EL display panel;
reference electrode means disposed on said thin-film EL display panel; and
means for deriving current having an amplitude equal to a difference between the current through said reference electrode and a readout current when reading out the memory state of said display panel.

4,225,808
SELECTIVE ILLUMINATION
Remo Saraceni, Philadelphia, Pa., assignor to Novitas, Inc., Santa Monica, Calif.
Filed Jun. 5, 1978, Ser. No. 912,442
Int. Cl.² H05B 37/02; F21P 3/00
U.S. Cl. 315—307   3 Claims

1. A method for illuminating items separated into at least one display group located in a display area comprising:
lighting the display area using dim background lighting;
sensing the approach of a person to said one display group;
switching automatically the lighting from the dim back-

ground level initially provided to a level of greater intensity upon the sensing;
sensing the departure of the person from the said one display group; and
returning automatically the lighting to its original dim background level upon the sensing of the departure,
whereby a display group may be highlighted with intense lighting when a person desirous of viewing the display is present and the display group may be maintained in a dimly lit manner when no person is present.

4,225,809
SIDE PINCUSHION CORRECTION CIRCUIT
Seiichi Ogawa, Tokyo; Yoshiaki Ohgawara, Inagi, and Kenichi Ohtsuka, Yokohama, all of Japan, assignors to Sony Corporation, Tokyo, Japan
Filed Apr. 6, 1979, Ser. No. 27,713
Claims priority, application Japan, Apr. 7, 1978, 53/41420
Int. Cl.² H01J 29/70
U.S. Cl. 315—371   16 Claims

1. A side pincushion distortion correction circuit for a cathode ray tube deflection apparatus including a horizontal deflection generator for generating a horizontal scanning current and a horizontal pulse during a retrace interval thereof and a vertical deflection generator for generating a vertical scanning current during a vertical scan interval and a parabolic wave signal synchronized therewith, comprising,
a horizontal deflection winding coupled to said horizontal deflection generator for accepting said horizontal scanning current;
an impedance circuit connected in series with said horizontal deflection winding;
controllable switch means including a controllable switch having a control electrode and a controlled current path connected in parallel with said impedance circuit;
switching signal generating means for generating a switching signal responsive to said horizontal pulse;
means for modulating the phase of said switching signal by said parabolic wave signal at the vertical rate;
means for supplying said switching signal to said control electrode of said controllable switch for operating said controllable switch during the latter half of said horizontal retrace interval;
means for progressively advancing said phase of said switching signal during a first portion of the vertical scan interval and for progressively retarding said phase during a second portion of the vertical scan interval for altering said scanning current in a manner to reduce pincushion distortion;
means for generating a brightness signal corresponding to the brightness of a reproduced picture; and
means for further modulating said phase of said switching signal in dependence on said brightness signal.

Figure 1.3   Page from *Patent Gazette* (U.S. Patent Office, September 30, 1980).

The statements are not newly expressed in this book. They are well recognized [3] and often lead to suggestions that faculty members and industry engineers participate in an exchange program in an attempt to remedy the problem of engineers being taught by faculty without industrial experience. Unfortunately, as one examines the problems confronting engineers from industry suddenly involved in course preparation, faculty terminating all projects for a year and then reestablishing them, and two families suffering dislocation, the exchange idea remains more rhetoric than activity.

There has been a full generation of engineers taught as the post World War II study recommended. If the increase in technically trained people is considered, Figure 1.1 shows that creativity as measured by patents to U.S. citizens has decreased. Perhaps ability to model a given product and to optimize it have increased but optimization typically results in incremental improvements. It is invention that often gives substantial improvement. Although it is not being suggested that the changes in the philosophy of engineering education and the makeup of engineering faculty are the only reasons for the present climate of invention, they are certainly among the important factors.

THE CHANGE IN INDUSTRY

Just as important as the change in *academé* was the change in industry during the past quarter century. This was a consequence of the evolution to the higher technology upon which the postwar escalation of our quality of life was based. The evolution included increased complexity of products, increased complexity of doing business, and increased concern with safety and environmental damage. Engineers find themselves concerned with record keeping, with meeting standards imposed by groups with authority, with defense of product liability suits, with customer service, and with extensive testing to verify quality and reliability. These are all proper activities but, nonetheless, they tend to distract the engineer from the pursuit of invention and subtly direct him toward the conservative approach of minor improvements on well-accepted design concepts. In brief, engineers do not have the freedom they once had. If the airplane, the automobile, and the rotary lawnmower had been invented in the climate of 1980 they probably would not have developed to the successful products they now are.

There is, of course, another change that took place in industry. This is the fact that most of our needs now must be met by recourse to complex manufacturing processes and precision tools. There are many areas of product development in which the need for expensive scientific equipment is so great that the independent inventor or the small company is not likely to be active there. This is unfortunate because the history of invention is replete with examples of individuals whose genius and persistence gave society a needed device or process. However, I am not saying that the individual inventor or the small company can find no areas for activity. There will always be problems that can be chiseled away from the complicated maze of system development and solved by brain power rather than hardware. It does mean that the situation whereby an individual working in a basement laboratory is at least as likely to develop a state-of-the-art product as engineers working in an industrial laboratory no longer exists.

THE CHANGE IN PEOPLE

For anyone whose business career has spanned the last quarter century the change in work ethic must be apparent. Perhaps because many of the workers of the 1950s remembered the depression of 1930 or perhaps because the relief of the war being over gave renewed dedication, the attention to the business of the employer was much more intense than it is now. Then, people seemed to accept their job as something they must do to compensate for the food they would consume, the clothes they would wear, and the home that would shelter them. Everyone seemed anxious to "get ahead." That gave status, a basis for being satisfied with oneself.

Now the mood seems to be that the job is much less the controlling factor of a person's life, that leisure is more important, that an employee should expect improvements in his job situation without the need to make unusual contributions to the employer. The prevalent mood suggests that the employer needs the worker more than the worker needs the employer.

The continued availability of good jobs lessened the motivation to work as intensely as invention demands to advance within the company or to spend time inventing as a freelance inventor. Adversity or personal insecurity has often served as a catalyst to inventors in the past and these threats have been lacking for the well-paid and sought-after engineer.

The change from work ethic to life ethic was accompanied by an increase availability of leisure activities. Both passive activities, such as television viewing, and active participation in sports have replaced the hours that many would-have-been inventors could have spent in basement workshops.

WHY INVENT?

The country needs renewed interest in invention and this must come, for the most part, from practicing engineers. The problems to be solved are too complicated for persons who do not understand physical phenomena and manufacturing methods. These problems have to do with the turn-around we must make from a wasteful, polluting society to a frugal people concerned with the quality of our environment. Yet "frugal" must not mean "deprived." The mood of the American people has never been to deprive themselves of anything they can possibly attain. We must return to a constant improvement in our quality of life but to do that we must find ways to use energy, material, and labor more efficiently.

The answer to the question "why invent?" is more satisfactorily given in terms of rewards to the inventor. Compensation is discussed in Chapter 13 and, in Chapter 4, a theory of invention is presented which helps to explain the feeling of elation and satisfaction that comes to an inventor. At this point, suffice it to say that invention and the consequent award of a patent is an effective way to give an engineer's career a lift no matter where he or she is between its beginning and retirement.

ARE THERE OPPORTUNITIES TO INVENT?

The purpose of discussing the present climate for invention is not to argue that opportunities to invent are any less numerous today than they have been in the past. The constant flow of new technology, new components, and new materials will give new ways to provide for our present needs and will allow newly identified needs to be addressed. The point is that the situation within which an engineer works has changed markedly and adjustment has not been made by educational institutions, by industry, or by the would-be inventors. Perhaps this book can make a significant contribution in returning creativity and invention to the engineering departments of U.S. manufacturers and to independent inventors. It describes the invention process with what is believed to be a viable theory of inventing, it gives advice concerning what you need to do to improve your ability to invent, and it presents structured methods that successful inventors use. Furthermore, you will find that the theory, advice, and methods support each other and confirm their reasonableness.

2

Examples of Inventions

Three examples of what might be called ordinary inventions will be discussed. The use of the word *ordinary* is not meant to demean these inventions in any way. All three are valuable to society and very rewarding to the inventor. However, they are not the landmark inventions we read about in history books—the steam engine, the reaper, the cotton gin, and so forth—nor are they the exceptional inventions which will be discussed shortly to give insight into the inventive process. The purpose of presenting these inventions is to convince you that significant contributions to productivity, safety, quality control, and the like can be made by inventions which some persons might describe as less than spectacular.

Liberal use will be made throughout this book of examples for which background information is known about the inventor or his invention. Much can be learned about inventing in this way. The main points of each case will be explicitly stated but you should look carefully for the more subtle implications of the stories. In the ones that immediately follow note that identifying the need is always the first step and in one of the inventions was essentially all that was required. Note also how each of these inventors came to discover the need by their work and because of their expertise. Note also the differences in age.

LOCKING THREAD CONSTRUCTION

The first of these inventions has to do with a way to retain nuts on machine screws. This need has been well recognized by others. There are split and tooth lockwashers, nuts with plastic inserts, screws that are out of round, and other ways too numerous to mention. However, in 1976, Horace D. Holmes [4], a 61-year-old inventor, developed still another way of locking a nut in place that has advantages over all others. In 1969 he had developed a "crimp nut" fastener system which is presently used by the automobile industry. That system requires an expensive tool which spins the nut onto a bolt and then squeezes it around the bolt threads to lock it in place. His latest invention requires no extra part, no extra manufacturing operations, no extra force to drive it before final tightening, and can be used with machine screws having rolled threads. Consider the reduction in labor and the improvement in quality control resulting from these features.

The nuts are made by a special tap which cuts the threads with a 30° wedge ramp at the root as shown in Figure 2.1. The screw thread has room to avoid touching the wedge until it begins to tighten because the outside diameter of a screw is always less than the root diameter of the corresponding nut. However, as the nut of this invention is tightened the screw threads are pulled against the wedge which centers the screw and eliminates the possibility of sideways motion between the screw and nut. A clamping action along the entire tip of the screw threads results, which increases the screw-to-nut thread friction and reorients the force vectors between the two to a more nearly radial direction. Force in the radial direction has no tendency to unseat a nut.

Holmes' invention required complete understanding of the mechanism by which a nut loosens on a machine screw. There is a tendency to gloss over such commonplace things or to be satisfied with a hazy idea of what happens. Invention requires a deeper understanding of physical phenomena than does other engineering activities.

Details of this construction are shown in Figure 2.2. The abstract of the patent (4,171,012) published in the October 16, 1979 (Vol. 987, No. 3) issue of the *Patent Gazette* [5] describes the invention as follows:

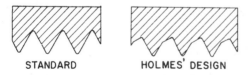

STANDARD   HOLMES' DESIGN

Figure 2.1 Threads of patent 4,171,012 compared with standard machine threads.

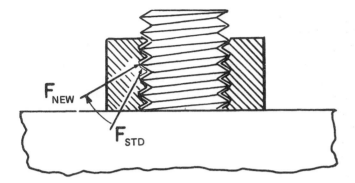

Figure 2.2 Nut with patented threads on standard screw.

A locking thread design which may be incorporated in various types of male and female threaded elements, for example, a bolt and nut, or a bolt and casting, forging or similar member having a threaded bore therein. The thread design may be of the Standard, i.e., American or Unified Standard, or buttress type and is free running until a predetermined magnitude of loading is applied thereto, at which time the locking action of the thread occurs so as to prevent relative lateral movement between the nut and bolt and hence positively resist loosening thereof under vibration and similar adverse operating conditions. The locking thread may be embodied on either one or both of the threaded elements, and will operate effectively when the locking thread is operatively associated with threaded members having conventional threads thereon.

Taps are available to produce these special threads and engineering evaluation is proceeding in many industries. In view of the billions of fasteners used each year, a locking nut that could be made simply by using a special tap to cut the threads would result in substantial industry-wide savings.

AN IMPROVED SOCKET WRENCH

In 1978 Peter M. Roberts [6] was awarded $1 million when a federal jury ruled that Sears, Roebuck & Co. had obtained rights to Roberts' patent fraudulently. When he was 18 years old, working as a clerk in a Sears store, Roberts invented a socket wrench which enables a mechanic to change sockets with one hand. It uses a fingertip-operated release for the spring-loaded ball which retains the socket on the rachet handle. A replacement socket is then easily pushed into place. This can be an important advantage if parts must be held by one hand during assembly, the socket operated with the other hand, and several different size nuts or bolts used. It is at least a convenience under any circumstance. According to court records, 25 million wrenches were sold at a profit of $44 million. According to newspaper accounts, the basis of the suit was that Roberts had been misled and sold rights to Sears for $10,000 because of statements by company representatives that there "wasn't much need for it and probably would not sell well . . ." but that the company was, in fact, changing its whole tool line to Roberts' design during negotiations.

Figure 2.3 shows the details of the construction. The patent (3,208,318) has only one claim and is the shortest disclosure of any referred to in this book. The invention is indeed a simple one but its commercial impact has been outstanding.

A GROUND FAULT INTERRUPTER

The last invention discussed in this section was made by Prof. Charles F. Dalziel of the University of California, Berkeley. He had a long and distinguished career there and is now Professor Emeritus of Electrical Engineering. He has published more than 110 technical papers many of which reported the results of experiments on the effects of electric shock on human subjects [7]. Dalziel used the information he had gathered to determine the limits of electric current above which muscular control is not possible (to escape from a hand-held conductor administering the shock). He combined this information, the use of improved magnetic materials which had become available in the early 1960s, and solid state electronics to produce a very compact sensor [8] which operates a circuit breaker to interrupt the electric current in milliseconds whenever the leakage current to ground from the conductors of a single-phase circuit is greater than 5 mA. Figure 2.4 shows a diagram of the system and the first claim of the patent (3,213,321). There has been a veritable deluge of patents on such sensors since Dalziel's. However, he is recognized internationally as the inventor of the residential ground fault interrupter.

Several of these devices are used in every new home. Typically, between 1.5 and 2 million homes (including apartments and mobile homes) are built in this country each year. Thus, the industrywide annual sales volume is estimated to be over $100 million.

Sept. 28, 1965   P. M. ROBERTS   **3,208,318**

QUICK RELEASE FOR SOCKET WRENCHES

Filed April 24. 1964

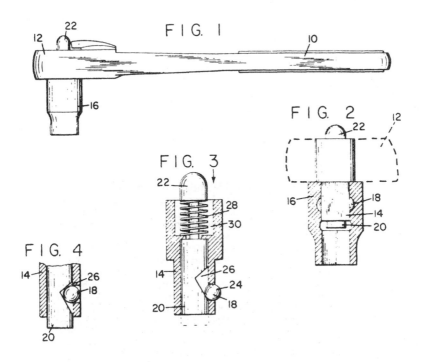

INVENTOR
PETER M. ROBERTS

BY   *Charles R. Fay*

ATTORNEY

Figure 2.3   First sheet of Roberts' patent (U.S. Patent Office).

3,213,321
MINIATURE DIFFERENTIAL CIRCUIT BREAKER
Charles F. Dalziel, 2240 Virginia St., Berkeley, Calif.
Filed May 31, 1963, Ser. No. 284,706
6 Claims.   (Cl. 317—18)

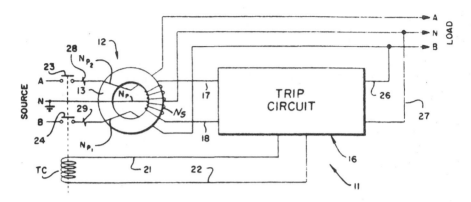

1. In a miniature differential circuit breaker primarily
intended for protecting human life by detecting a high
resistance line-to-ground short circuit or an abnormal
leakage current to ground from conductors forming at
least one path in an electrical circuit connected to a source
of electricity, current interrupting breaker contacts
adapted to be connected into the conductors, one of the
conductors of said electrical circuit being grounded at
the source of electricity, a differential transformer having
a single magnetic core of high permeability material and
having at least two conductors of said electrical circuit
passing through the single core so as to provide two or
more primary windings having at least one turn, the differential transformer also having a multiple turn secondary winding having sufficient turns so that the minimum
trip current is below 50 milliamperes, the primary windings being arranged so that under normal current conditions the total magnetomotive force produced in the single
magnetic core is balanced out so that the net magnetic
flux in the core is zero and under current-flowing-to-
ground conditions a net flux results which produces a
voltage in the secondary winding of the differential transformer, highly sensitive non-magnetic switching means
connected to the secondary winding of the differential
transformer and being operated when the current flowing to ground reaches a pre-selected minimum value below 50 milliamperes and means operated by said switching means to interrupt the electrical circuit by operating
the current interrupting contacts of the differential circuit
breaker.

Figure 2.4   Entry in *Patent Gazette,* Oct. 19, 1965 (U.S. Patent Office).

THE INDEPENDENT INVENTOR
AND THE EMPLOYED INVENTOR

These three inventions were made by independent inventors. Holmes had made previous inventions in the field of fasteners, and Dalziel was completing, in appropriate fashion, a lifetime concern for avoidance of injury and death by electricity. Roberts, on the other hand, seems to have hit upon his invention by the accident of having daily contact with the wrenches he improved and recognizing the need for that improvement. One can speculate that the need could have been exposed by a customer complaint. No matter, Roberts responded to the need regardless of how it became obvious to him.

This book is meant to promote inventions by persons in circumstances as diverse as those of these three inventors. It is also meant to serve persons in industry. Usually inventions in industry do not begin with an avowed intention to invent something. Rather, the designer recognizes a need as a development project progresses and seeks some way to satisfy that need. In industry, a simple invention such as those just described would be of great importance to the employer.

The capable designer strives to produce the best design that will meet the specifications. Just what is meant by "best" will depend upon many factors. For the moment, all you need to accept is the fact that rarely, if ever, does a designer happen upon the best solution to a design problem when he or she sketches the first system or device that seems workable. This solution should be evaluated and then set aside while attempts are made to create alternative solutions. Hopefully, one of the solutions will be close to the elusive "best" design.

Expired patents are very valuable sources of ideas for alternative designs. You should not feel defensive about basing your design on an expired patent. The very reason an inventor is given exclusive rights to his invention is that it will be available for anyone to copy after 17 years. This procedure has an added advantage in that if you copy an expired patent closely you need not worry about infringing an unexpired patent that you may have overlooked or one that is about to issue. Some very successful products have been developed based on expired patents. Of course, this technique does have the disadvantage of allowing others to copy your product.

Unexpired patents are also a valuable source of ideas which can be used free of infringement if done skillfully. Care must be exercised to ensure that the design concepts used do not infringe any of the claims [9]. Often, the evolution that takes place during a product development will change the design enough from the original concepts to avoid infringement, but the final design should be evaluated by a patent lawyer.

In seeking ideas for alternative designs you should not overlook publications of professional societies, trade journals, and manufacturers' application literature. You will find that other engineers who have been confronted with similar design problems have eagerly reported their solutions to build their professional stature. However, you need to be concerned about the patent situation of any design you copy from publications such as those named. The fact that a design is described in the literature does not imply that anyone may use it. It may be the subject of a patent or of a patent application.

After you have searched all the available sources for alternative designs of your proposed product, you may decide that a basically new approach is desirable. You may wish to generate a design that is superior to those you have found, or develop a new design which would permit the patent protection you deem necessary, or simply avoid infringement of a competitive patent. Any of these is proper motivation for you to try to invent a new device, system, or process to meet the perceived need.

TWO KINDS OF INVENTION PROBLEMS

Most inventions result from recognition of a need and an attempt to fill that need. However, there is another class of inventions that results from the recognition of a new phenomenon, material, process, or device which strikes the observer as being useful in some unexpected way. For example, Robert Lester [10], an independent inventor, looked at his liquid crystal watch while waiting to use a copy machine. The thought struck him that the copier should be able to reproduce the numbers on the watch face. A quick experiment showed that the numbers appeared on the copy paper with good quality. Extending this, he visualized an array upon which letters, numbers, and symbols could be made to appear by the same process used in the watch and then copied by xerography. He expects his typewriter to use much less energy than an conventional typewriter, last a lifetime, and possibly cost less.

3
Learning from Great Inventors of the Past

One of the objectives of this chapter is to convince you that you should not be overawed at the idea of being an inventor. Most inventions are not great, complicated discoveries that appear full blown in the mind of the inventor. Even the important, famous ones often involve a series of small steps forward in humanity's ability to do useful things. Also, usually the new or improved product as we know it is the integration of many of these small steps, some of which were made by other contributors. To show that even exceptional inventions are the result of desire and much work, a few landmark inventions will be presented.

THE STEAM ENGINE

The statement usually heard is that the steam engine was invented by James Watt. Watt's description [11] of the moment of invention follows:

> It was in the Green of Glasgow. I had gone to take a walk on a fine Sabbath afternoon. I had entered the Green by the gate at the foot of Charlotte Street—had passed the old washing-house. I was thinking upon the engine at the time, and had gone as far as the Herd's-house, when the idea came into my mind, that as steam was an elastic body it would rush into a vacuum, and if a communication was made between the cylinder and an exhausted vessel, it would rush into it, and might be there condensed without cooling the cylinder. I then saw that I must get quit of the condensed steam and injection water, if I used a jet as in Newcomen's engine. Two ways of doing

this occurred to me. First, the water might be run off by a descending pipe, if an offlet could be got at the depth of 35 or 36 feet, and any air might be extracted by a small pump; the second was to make the pump large enough to extract both water and air. I had not walked farther than the Golf-house when the whole thing was arranged in my mind.

Steam engines had been used for at least a hundred years. Watt had worked diligently for two years prior to this Sunday reconditioning a used engine which had been purchased by the University of Glasgow, where he was employed. The previous models had required that the steam be cooled in the engine cylinder. Watt's great contribution was to exhaust the steam into an auxiliary chamber, the "condenser," for cooling. This allowed the engine cylinder to be insulated and to remain hot.

His engine used approximately one-fourth as much fuel as the Newcomen engine to do the same work. A great contribution indeed; but it resulted merely from relieving the engine cylinder of the need to be heated and then cooled during each cycle of operation. The key to his invention was a hunch that steam would act as a liquid and would flow from the cylinder to another, cooler vessel.

In words taken from Watt's British patent (No. 913 granted 1769), "My method of lessening the consumption of steam, and consequently fuel in fire engines consists of the following principles:

First, that vessell in which the powers of steam are to be employed to work the engine, which is called the cylinder in common fire engines, and which I call the steam vessell, must during the whole time the engine is at work be kept as hot as the steam that enters it, first, by enclosing it in a case of wood or any other materials that transmit heat slowly; secondly, by surrounding it with steam or other heated bodies; and thirdly, by suffering neither water or any other substance colder than the steam to enter or touch it during that time.

Secondly, in engines that are to be worked wholly or partially by condensation of steam, the steam is to be condensed in vessels distinct from the steam vessels or cylinders, although occasionally communicating with them. These vessels I call condensers, and whilst the engines are working, these condensers ought at least to be kept as cold as the air in the neighbourhood of the engines by application of water or other cold bodies.

Five more "principles" are given in the patent concerning such details as obtaining circular motion, the use of animal fat for seals, and the use of steam above atmospheric pressure to move the piston.

Like most inventors of note, Watt did not stop with the success of his first invention. He spent the rest of his life improving components to develop higher pressure, more efficient engines.

XEROGRAPHY

Chester Carlson invented the process now known as xerography. His story is especially informative [12]. As a boy he got a job working for a printer. Later, after graduating with a degree in physics during the 1930 Depression, he worked briefly as a research engineer at Bell Telephone Laboratories in New York City, then for a patent attorney, and then for the electronics firm P. R. Mallory & Co. While at the latter position he studied law at night, earned a law degree, and eventually was promoted to manager of Mallory's patent department. This fortuitous combination of experiences taught him the difficulty of getting words into clear hard copy as well as the need for a convenient process to duplicate printed documents such as patents. His physics background directed him to do library research on imaging processes and he found accounts of work done by a Hungarian physicist, Paul Selenyi, on producing images by electrostatic processes. He chose to do research on this phenomenon because it was new and unexplored. There was no single rush of an idea that made clear what he should do. His invention was the result of careful research which taught him what he needed to know about the process. He filed his first patent application in October 1937. However, years of further research were needed to refine the process, first with another young physicist and then with the research organization Battelle Memorial Institute. It was not until 1957, 21 years after Carlson had made the first reproduction by xerography, that a useful office copier based upon his invention was available.

Figure 3.1 shows the steps of his process. The patent (2,297,691) describes the figures on this page as follows:

Figure 1 is a section through a photographic plate according to my invention and illustrates a preferred method of applying an electric charge to it preparatory to photographic exposure;

Figures 2, 2a, and 2b illustrate three methods of photographically exposing the plate;

Figures 3 and 4 show a method of developing the electrostatic latent image produced on the plate by the preceding steps;

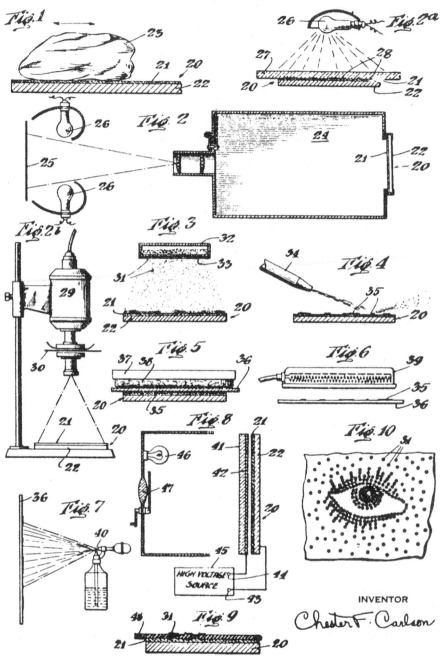

Figure 3.1 Carlson's copying system (U.S. Patent Office).

Figure 5 shows a method of transferring the image to a sheet of suitable material such as paper;

Figures 6 and 7 illustrate methods of fixing the image onto the sheet;

Figure 8 illustrates a modified means for charging and exposing the photographic plate;

Figure 9 shows another method of developing the image; and

Figure 10 is an enlargement of a half-tone produced by the process.

This development required dedication to and faith in the eventual successful outcome of his invention far in excess of what most persons are willing to give. The honors he received and his financial reward were spectacular. At 14 he was the sole support of his ailing parents, but by the time of his death at the age of 62 it is estimated that he had given away approximately $100 million to foundations and charities. This amazing increase in wealth resulted from his development of a totally new copying process.

THE NEGATIVE FEEDBACK AMPLIFIER

Another landmark invention was that of the negative feedback amplifier invented by Harold S. Black on August 6, 1927 [13]. Black started working for Bell Laboratories in 1921, immediately after graduation with a B.S.E.E. degree. He was not the usual type of employee. For example, to learn about the company and the telephone business, Black began coming in on Sundays to read through a collection of important memoranda which the company kept in file. He started this document study with the 1898 collection and states that by the time he reached the 1921 file he knew the technical problems facing the company.

Black's advantage was that he seemed to have insight into how the telephone business would grow to meet the needs of this country. While the immediate problem was to reduce the distortion of push-pull vacuum tube amplifiers carrying three channels over lines extending 1000 miles, Black made preliminary estimates of how good amplifiers would need to be to handle 3000 channels over 4000 miles. He found the requirements far beyond the state-of-the-art.

Western Electric research people were also aware of the need to reduce amplifier distortion for the more modest telephone network they were considering. They and Black tried to reduce distortion by improving the linearity of the vacuum tube. His important forward step, however, was taken when he directed his attention to the amplifier as a whole. He realized that his objective was to remove all distortion from the amplifier output. He writes, "In doing this, I was accepting an imperfect amplifier and regarding its output as composed of what was wanted plus what was not wanted. I considered what was not wanted to be distortion . . . and I asked myself how to isolate and eliminate this distortion. I immediately observed that by reducing the output to the same amplitude as the input and subtracting one from the other only the distortion would remain. The distortion could then be amplified in a separate amplifier and used to cancel out the distortion in the original amplifier output" [13, p. 58]. This line of reasoning occurred at 2:00 AM on March 16, 1923, after returning home from an American Institute of Electrical Engineers meeting in New York City where Charles Steinmetz gave a lecture that impressed and evidently inspired Black with its clarity and logic. The next day Black sketched two embodiments of the scheme and set them up in the laboratory thereby inventing the feed forward amplifier.

This invention reduced the distortion by 40dB. However, it required such precise balance and subtraction that it was difficult to maintain the advantage that was theoretically possible. Black continued this work, but every circuit he devised turned out to be far too complicated. He sought simplicity and perfection.

Then on the morning of August 2, 1927 the concept of the negative feedback came to him in a flash while he was crossing the Hudson River on the Lackawanna Ferry on the way to work. After years of intense effort he had suddenly realized that if the amplifier output was fed back to the input, in reverse phase, the means of cancelling out the distortion in the output would be realized. He sketched a simple diagram of a negative feedback amplifier on the copy of a newspaper he had in hand and derived the equation for amplification with feedback. On December 29, 1927, using typical input signals covering a frequency band extending from 4 to 45 kHz, a reduction of distortion of 100,000 to 1 (50 dB) was obtained in a single amplifier. Figure 3.2 shows the improvement in linearity realized by various degrees of feedback.

THE DUPLEXER

The last landmark invention we will discuss is described in a delightful little book of Robert Morris Page [14]. The basis of radar, i.e., the reflection of electromagnetic waves, was discovered by accident in 1922 as A. H. Taylor and L. C. Young were studying high-frequency radio communication by transmitting on one side of the Anacostia River in Washing-

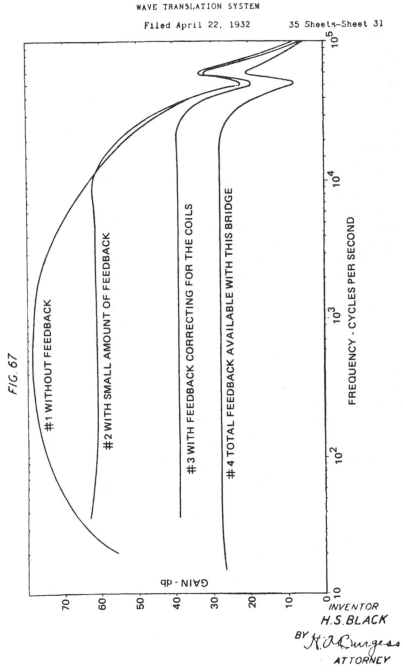

Figure 3.2 Curves showing constant gain over wider frequency limits provided by feedback (U.S. Patent Office).

ton, D.C. and receiving on the other side. A river steamer interfered with the experiment. It was not surprising that there would be interference when the boat cut the line of sight between the two antennas. However, the signal periodically strengthened and waned as the boat approached and as it continued up the river. That variation was not expected. In 1930 there was an accidental aircraft detection by a shortwave transmitter emitting a steady tone which was being received by a shortwave receiver several miles away. Thus, as the danger of World War II approached, the use of reflected radio waves to detect aircraft and ships was established.

There were many major scientific and engineering problems to solve in the development of the radar system. One of the most challenging began as a mechanical engineering problem, namely, how can the transmitting antenna and the receiving antenna be so accurately controlled that they are aimed at essentially the same point in space 10 miles, 15 miles, or even farther away? They could not be, so a new approach had to be investigated, namely, using the same antenna for transmitting and receiving. But this created a new problem. How can one antenna be switched from a transmitter which just sent out a pulse at the rate of a megawatt to a receiver which will detect a reflection of a milliwatt?

Page accomplished that by permanent connections from the transmitter and the receiver to the antenna by means of quarter-wavelength lines. Figure 33 of his book is reproduced here as Figure 3.3 for those who have had previous work in transmission lines. But his story can be appreciated by all. The important points to be considered are well expressed by an excerpt from Page's book.*

The duplexer just described was first built and tested in the summer of 1936. It worked perfectly the first time it was tried, and its success leads me to something that has been a significant part of all my research. Back in Chapter 1, I said that I believed that some new ideas appear as if by accident, containing something more than existed in prior knowledge. Dr. Taylor, head of the Radio Division, had said we ought to try using a single antenna, since transmitter and receiver did not have to be operating simultaneously. I had said it was impossible, because, if for no other reason, the receiver would be burned up by the transmitter. Leo Young, my immediate supervisor, then asked me if I could not use some arrangement of

*Figure and excerpt from *The Origin of Radar* by Robert Morris page, illustrated by Kenneth Cook. Copyright © 1962 by Educational Services Incorporated. Reprinted by permission of Doubleday and Company, Inc.

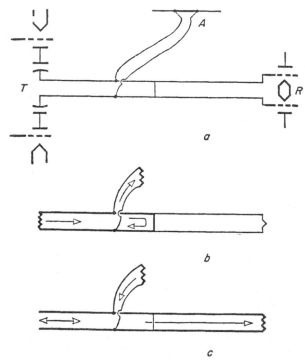

Fig. 33. The first radar duplexer switched with electrons in the grid circuit of the receiver input tubes.

(a) A half wavelength, two-wire line is shorted at the center, terminated in the transmitter output at T and in the receiver input at R. The antenna, A, is connected through a transmission line to the proper impedance point on the transmitter side of the half-wave line.

(b) During transmission of the pulse, the receiver resistance is very low, so very little of the transmitter power gets to the receiver, and the transmitter is matched to the antenna for good power transfer.

(c) During reception the receiver resistance is high and the transmitter resistance is much higher, so the two quarter-wave lines are tightly coupled through the common shorting bar at the center. The antenna is matched through the resonant lines for good signal transfer to the receiver.

Figure 3.3 An electric circuit which channels energy from the transmitter (on left) to the antenna but channels energy received by antenna to receiver (on right). (From *The Origin of Radar* by Robert Morris Page, illustrated by Kenneth Cook. Copyright © 1962 by Educational Services Incorporated. Reprinted by permission of Doubleday and Company, Inc.)

tuned circuits with spark gaps to protect the receiver. It was obvious that I had to try something.

I did not understand impedance inversion at the time, although I had heard that there was such a thing. I knew, however, that the input impedance of vacuum tubes was high when grids were negative, and low when they were positive. So I figured that if they were matched by a high impedance circuit for the negative condition, when they were driven positive they would be so badly mismatched that very little of the input energy would be absorbed. Spark gaps should therefore not be necessary. Then the transmitter coupling to the antenna would have to be accomplished with another circuit, which would not be rendered inefficient by the receiver loading during the transmitted pulse. The arrangement of two quarter-wave lines connected back to back as I have described was purely the result of a hunch. I had no intellectual idea whether it would work or not, for I did not understand how it worked, even after it was successful. I did have a subjective conviction that it would work. This conviction, or faith as some would call it, was so strong that when it proved successful I was more elated than surprised. It was not until many years afterward, when several other people were claiming invention of the radar duplexer and everyone had a different explanation of its operation, that I was forced to give a rigorous explanation of how it did work. Then for the first time I think I began really to understand it. Then it appeared that the original form in which I first tried it was the most simple, most direct, and, for the frequencies used, most efficient design I could have made. It was referred to by patent attorneys as one of those rare cases of "flash of genius" when something really new and basic appears. But in all sincerity I can take no personal credit for it, because I did not create it. I only followed a "hunch," or, as I prefer to call it, an inspiration, in which the completed configuration appeared in my imagination without an understanding of how it worked, but with a feeling of great confidence that it would work. It was as if a source of knowledge out of this world had momentarily been opened to me, and I was guided by it. This is but one of many such experiences that have marked my professional career. Do you wonder that my faith in Divine Providence is both profound and precious?

SUMMARY

The common thread that runs through all four of the above examples is the dedication and intensity with which Watt, Carlson, Black, and Page pursued their goals. Beyond that, and except for a common characteristic I will discuss later, many differences appear. For example, note how each learned of the need for his invention. The intolerable inefficiency of the steam engine was common knowledge; Page's supervisor defined the need for a switching device and even suggested a system for consideration; however, both Carlson and Black looked to the needs of the future in the light of then existing state-of-the-art to decide that entirely new concepts were necessary. In the matter of approach, Carlson's was methodical and scientific. His biographies tell of the exciting moment when success in producing a copy was first achieved, but no dramatic moment of insight when the process flashed in his mind's eye as was true of Watt and Black. In fact, Carlson is quoted as saying that ideas came slowly. Finally, Page's invention is perhaps the most awe inspiring. The analysis of electrical transmission lines was not well developed at that time. In his book he presents a heuristic explanation of electron movement in a shorted and an open quarter-wavelength line but this was developed after the invention. His story more than the others shows the heights to which the human mind can ascend.

The other common characteristic to which I referred was stated by Black when he said he sought a solution that was simple and perfect. This is a distinguishing characteristic of all worthy inventions, even ones of much less importance than those discussed here. I use the word *elegant* to describe what Black meant. That word is no less difficult to define when

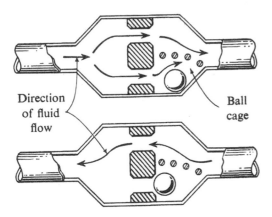

Figure 3.4 Heat pump valve. (From W. H. Middendorf, *Engineering Design,* Allyn and Bacon, Boston, 1969. Copyright W. H. Middendorf.)

applied to invention but it is easy to recognize. Remember the definition as the word appears often throughout this book.

The meaning of *elegant* is best taught by an example. A number of years ago I attended a talk given by a manufacturer of a line of heat pumps which was commercially available. During the lecture the speaker mentioned that twice as much fluid was used when the pump was cooling than when it was heating. Also, it was clear that the fluid is pumped in opposite directions during heating and cooling. Since a listener thinks faster than a speaker talks, the thought immediately came to mind that a monitoring unit was involved with transducers to respond to the direction of flow and a control system to open and close a valve. I could see most of the components in my mind's eye. Such a system could have undoubtedly been made to do the job. However, later in the talk the speaker showed a slide similar to Figure 3.4. This made quite an impression. The use of the change of direction of the fluid to seat or open a ball valve is indeed elegant. The massive system that came to my mind was an example of what is done so often, namely, overblow the problem rather than solve it. My system was not "elegant."

4

Theories of Creativity

Creativity is common to a great number of endeavors, both scientific and artistic. The flash of creative insight experienced by an engineer engaged in invention is very similar to that felt by the poet as he finds just the right word, the mathematician in discovering the solution to a difficult proof, the musician engaged in composition. The methods used to stimulate creativity are also similar for all disciplines. Psychologists have been interested in the creative process, the creative personality, and methods of inducing creativity since the time of Freud. Out of this interest have arisen theories of creativity, and a composite personality of the creative person.

Before any studies were done by psychologists the only material available on the subject of creativity was of an autobiographical nature. The important information contained in these studies is that creative ideas come from outside the realm of conscious thought but that creation is not merely inexplicable inspiration. It involves a lot of effort and applied skill on the conscious level. [15] The following are thoughts of Mozart, Tchaikovsky, and Poincaré on the subject of their own creativity:

> When I am, . . . entirely alone, and of good cheer; . . . it is on such occasions that my ideas flow best and most abundantly. Whence and how they come I know not; nor can I force them. Those pleasures that please me I retain in memory, and am accustomed, I have been told, to hum to myself. . . .
>
> All this fires my soul, and provided I am not disturbed, my subject enlarges itself, becomes methodized and defined, and the

whole, though it be long, stands almost complete and finished in my mind, so that I can survey it, like a fine picture or beautiful statue . . .

What a delight this is I cannot tell: All this inventing, this producing, takes place in a pleasing lively dream" [15, p. 55].

Generally speaking, the germ of a future composition comes suddenly and unexpectedly. If the soil is ready—that is to say, if the disposition for work is there—it takes root with extraordinary force . . . and finally blossoms. . . . The great difficulty is that the germ must appear at a favorable moment, the rest goes of itself. It would be vain to try to put into words that immeasurable sense of bliss which comes over me directly a new idea awakens in me and begins to assume a definite form. I forget everything and behave like a madman. . . .

Dreadful indeed are interruptions. Sometimes they break the thread of inspiration for a considerable time. . . . In such cases cool headwork and technical knowledge have come to my aid. . . . It is a great thing if the main ideas and general outline of a work come without any racking of brains, as the result of that supernatural and inexplicable force we call inspiration" [15, pp. 57-58].

Most striking at first is the appearance of sudden illumination, a manifest sign of long, unconscious prior work. The role of this unconscious work in mathematical invention appears to me incontestable, and traces of it would be found in other cases where it is less evident. Often when one works hard at a difficult question, nothing good is accomplished at the first attack. Then one takes a rest, . . . and sits down anew to the work. During the first half hour, as before, nothing is found, and then all of a sudden the decisive idea presents itself to the mind. It might be said that the conscious work has been more fruitful because it has been interrupted and the rest has given back to the mind its force and freshness. But it is more probable that this rest has been filled out with unconscious work and that the result of this work has afterward revealed itself to the geometer" [15, p. 83].

SELECTED THEORIES

Psychologists have only very recently turned their research efforts toward development of a theory of creativity, although some speculation went on previously. All theories are not based on the idea of inspiration arising out of the subconscious and preconscious mind as a result of some sort of transfer of a problem from the consciousness to lower mental states where it is mulled over until magically solved; although, this is the theory to which I subscribe and which is presented later. E. W. Sinnot [15] claims that, although some new ideas appear to arise almost spontaneously, there is a second major method, that of creativity by direct frontal assault. In this method the widest possible array of facts and ideas are collected and then a search is made for previously unseen relationships between these facts and ideas. Much of Edison's work was done in this manner. He often collected little known inventions of others and assembled them into inventions of his own. Sinnot also suggests that creativity is related to the ability to pick out important facts and ideas from the vast collection stored in the mind. This is because of the mind's organization of information into categories.

At the extremes of psychology are the stimuli-response theorists and the cognitive theorists. Both these schools of psychology have developed theories of creativity. The stimuli-response theory suggests that creativity is the formation of associations between stimuli and responses which are not normally associated. Creative people are particularly skillful at connecting aspects of their environment which on the basis of experience do not seem to belong together.

The cognitive theory holds that the creative individual organizes everything into categories on a subconscious level as do all individuals, but the creative person's categories tend toward divergence rather than convergence, and as a result of this divergence the creative person can recognize relationships which would not otherwise be apparent [15].

Still another theory of creativity is advanced by C. R. Rogers. Rogers states that creativity is the emergence of a new idea caused by the interaction between a unique individual and the events, people, and circumstances of his life [15]. As already indicated by every one of the inventions discussed to this point, each person's special skills and/or circumstances certainly do influence the contributions he or she can make. According to this theory, certain conditions must be present in the creative person and certain conditions must be present in the creative person's environment. Their coincidence is somewhat a matter of luck but the more talented the individual and more varied his or her experience the more likely the coincidence will occur.

The theory of creativity that seems most plausible and useful to me is taken from a book by Koestler [16]. It seems plausible because it conforms to my personal experiences and useful because it clearly indicates what must be done to increase creativity.

A basic tenet of the theory is that all creativity has the common char-

32

acteristic that a relationship is seen to exist between two entities which are not previously recognized as being connected. This is true even if the creative act is merely the construction of a good joke. The thought progress can be shown diagrammatically by Figure 4.1. The vertical plane represents an area of thought and all the ideas one would normally associate with that area of thought. As our mind scans the limits of that plane there are no surprises; we might even say that any train of thought contained therein is "common sense" and familiar to those "skilled in the art." However, suppose there is another plane of thought not obviously connected to the first to which our mind might jump and in doing so get the solution. This is represented by the horizontal plane. Koestler calls this jump "bisociation."

Kestin [17], following Koestler, gives a simple but excellent example of this moment of insight. As a boy he was challenged by the problem of drawing a right triangle when given two lines; one being the hypotenuse, C, and the other, H, being the distance from the right angle perpendicular to the hypotenuse as shown in Figure 4.2.

His first approach was to draw a right angle having sides of indefinite length. Then he attempted to visualize the hypotenuse sliding with ends attached to the right angle's sides until a position is reached which would give the correct length of H and the required relationship of it to C. Numerous trials could establish the triangle—at least to a close approximation—but that was not an acceptable solution. The next morning the problem appeared on a quiz. Kestin visualized it in a different orientation as shown on the left in Figure 4.3. He had seen this orientation before, related to a theorem of geometry that states that the angle sub-

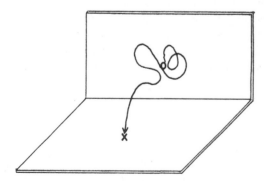

Figure 4.1 Intersecting planes of thought.

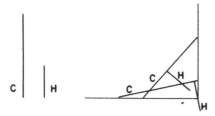

Figure 4.2 Kestin's first attempt to solve the right triangle problem.

tended on the diameter by a point on the circle is a right angle. The reorientation resulted in the jump to recall a theorem that did not occur to him during the previous evening's study. However, once the connection was made the exact solution was easily obtained; in fact, that construction even defines the maximum height the right triangle can have, namely, $H_{max} = C/2$. For the young Kestin this was an invention.

In his extensive treatment of this theory of creativity [16], Koestler cites the invention of the printing press as an example of this moment of insight, this association between two planes of thought which in this case existed but remained unassociated for hundreds of years. He states that letters testify that Gutenberg had long engaged in many attempts to improve the old art of printing. The art of making playing cards and pictures of saints by rubbing cards on engraved woodblocks was well known and the art of making coins by striking a die dated back many centuries. However, these skills were not adequate for printing a book. In that application the method needed to apply more composition with each impression and the pressure applied to the paper with precision.

Gutenberg took part in the wine harvest. He wrote, "I watched the wine flowing and going back from the effect to the cause I studied the power of this press which nothing could resist" [16, p. 123]. At this mo-

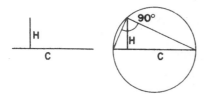

Figure 4.3 The perfect solution to the right triangle problem.

ment it occurred to him that the same, steady pressure might be used to press type to paper and then remove the type straight away from the paper to avoid smudging. Thus, the result of having a person skilled in printing, who had recognized the need to improve the process, witness an operation that seems totally disassociated, wine making, resulted in one of the most important inventions of all time. It does not matter that the process had been developed in China sometime earlier. For Gutenberg it was a totally creative act and for Western civilization it provided a new era for information storage and universal distribution.

The intersecting planes of thought shown in Figure 4.1 take on clearer meaning if the concepts involved in Gutenberg's invention are identified on each plane. This has been done in Figure 4.4. There are some factors which are common to wine making and printing by press. These define the line of intersection between the two planes. If such common requirements are identified as the inventor seeks a solution the bisociation usually follows. In both planes there are also many other things or concepts which are pertinent only to one or the other. These lie far off the line of intersection as shown by the few named; paper, ink, grapes, bottles, and kegs. As Gutenberg viewed the action of the wine press the insight that the same basic mechanism would be appropriate for a printing press was accomplished instantly.

Perhaps it is a bit difficult to fully appreciate how great a jump in thought was required by Gutenberg's invention. Remember that there were no books, pictures, or the many other ways we have to transmit in-

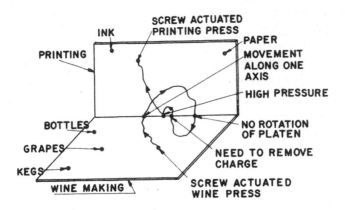

Figure 4.4 Intersecting planes of wine making and printing.

formation. It is conceivable that Gutenberg had never seen a screw before or, even more likely, that he was completely unaware of its mechanical advantage before he saw the grape juice gush from the wire press.

It should also be mentioned that Gutenberg's work did not stop at that moment of bisociation. That was the beginning of intense effort to improve the ink and to develop movable type suitable for the press [18]. This activity probably involved subsequent lesser inventions or it may have been skillful engineering and research. It is not unusual for the insight to an elegant solution to act as the motivation for prolonged and intense effort to complete the invention.

AN ORDINARY EXAMPLE

It is appropriate that I present an example more closely related to the type of problem likely to be encountered in product design. In the early 1950s I was involved in a product development that resulted in my first invention. The overall problem was to develop a residential-type circuit breaker which would occupy only half the space of the one then produced by the client company. The width of the unit in production was 1 in.; the new one was to have two devices in a 1-in., molded plastic case.

Circuit breakers are a very appropriate product to consider because overall they involve most of the engineering disciplines. Figure 4.5 shows one of the many designs used. The case of thermosetting plastic involves chemical engineering, the silver-tungsten contacts and the special alloys which have high mechanical strength as well as high electrical conductivity involve metallurgical engineering, the manual control and displacement amplifier involve mechanical engineering, and the overcurrent sensor combines electrical and mechanical engineering problems. The device must be inexpensive and yet reliably interrupt the electric current when it exceeds the current rating of the wire which the circuit breaker is designed to protect.

All circuit breaker mechanisms involve a latch (parts 46 and 47 in Figure 4.5) which disengages in response to excessive values of current through the device. To provide a means of tripping on a modest overload, most designs have a length of bimetal (17, two metals of dissimilar thermal expansion side by side) which is heated by the current and the resulting movement is used to disengage the latch. However, that process is too slow for high-current overloads and so magnetic forces are used to cause disengagement. The decision was made to have the design of the

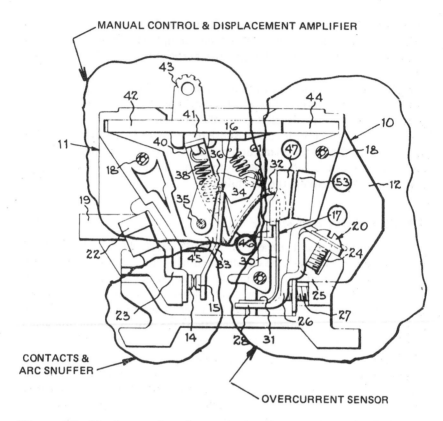

MANUAL CONTROL & DISPLACEMENT AMPLIFIER

CONTACTS &
ARC SNUFFER

OVERCURRENT SENSOR

Figure 4.5 The interacting elements of a circuit breaker (U.S. Patent Office).

miniature device similar to the one already in production to benefit from the manufacturing processes and skills which had been developed for it.

The design problems encountered in reducing the size of components were minor until the magnetic circuit was considered. The unit in production used two U-shaped steel pieces, one, the latch (47), which was welded to the bimetal, and the other, the armature (53), which was held stationary in depressions molded in the circuit breaker housing. A magnetic field was produced in the two steel pieces and the intervening air gaps by the electric current which passed through the bimetal and then to a flexible conductor as shown in Figure 4.6. The latch was pulled into the armature by the magnetic field during high-current overloads.

The armature was made in a U shape so as to present a short air gap when the latch is in position to hold the trip mechanism and yet not to hinder the latch movement as the bimetal continued to flex due to the current during the arcing period which usually follows separation of the contacts. If bimetal is restrained from moving when it is hot its internal forces can exceed the yield strength and distort it. This changes the circuit breaker calibration.

It is easily seen from Figure 4.6 that the magnetic circuit pieces, 47 and 53, required four thickness of material plus clearance to ensure unimpeded motion. The magnetic pieces needed to be 16-gage steel (1/16 in. thick) to provide a sufficiently low reluctance path for the magnetic flux during short circuit. Thus, with a 1/32-in. clearance on each side, a total of 5/16 in. would be required to accomodate the magnetic pieces. The minimum thickness of the plastic sides and center piece for proper curing was to be 1/16 in. each, leaving a mere 3/32 in. within the total 1/2-in. width of the circuit breaker for the bimetal to which the latch was to be welded. This was not sufficient width for the bimetal. Thus, the problem was well defined. The design sought was one which would pull the latch away from the trip mechanism by magnetic force during high-current overloads but then immediately allow continued movement in the same

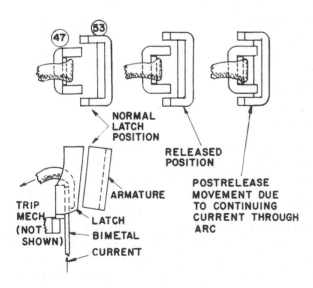

Figure 4.6 The electromagnetic tripping unit of a circuit breaker.

direction as the bimetal continued to flex. Furthermore, it must allow the bimetal to be at least 3/16 in. wide.

A number of rather prosaic ideas occurred. For example, the armature could be flat and mounted on a leaf spring. This would eliminate two thickness of metal and allow the bimetal continued movement during the arcing period by bending the thin spring. Abutments would be used in the plastic housing to restrain the armature from moving to the latch rather than the other way around. This and other ideas were rejected for various reasons and the vexing problem of designing the magnetic circuit held up development for several weeks. Then, one Saturday morning when I was alone in the office and deeply involved with the problem the bisociation took place (although I did not know of Koestler's theory at the time). I began by comparing the sought after magnetic circuit to typical design of a current relay. There, a light armature carrying a contact is pulled toward a stationary contact by a strong electromagnet. The thought occurred to me that if I were to suddenly block the movement of the armature the heavier electromagnet would be pulled to it. I used my hand against the edge of the desk to simulate the action. The straight fingers represented the armature, cupped palm the latch, and the knuckles a hinge. All that was needed after that moment of insight was to adapt the dimensions to the small size to fit the circuit breaker. As shown in Figure 4.7, the need for the U-shaped armature was eliminated. The flat armature is mounted on the latch by a hinge and is carried unimpeded in the direction the bimetal must flex. This construction allowed the bimetal to be 3/16 in. wide.

The unit was used for many years in the client's product. It was convenient to manufacture and reliable.

JANUSIAN THINKING

A kind of creative leap that has been recognized is named after Janus the Roman god, whose two faces permitted him to look in opposite directions at once. "Janusian thinking" consists of actively conceiving two or more opposite or antithetical concepts, ideas, or images simultaneously, both as existing side by side and equally operative or equally true. An account written by Einstein in 1919 [19] describing his development of the general relativity theory gives an example of this type of thinking. He drew an analogy between the need for relative motion between a magnetic field and a conductor if electromagnetic induction is to take place (the conductor must "cut" the magnetic flux lines for voltage to be induced)

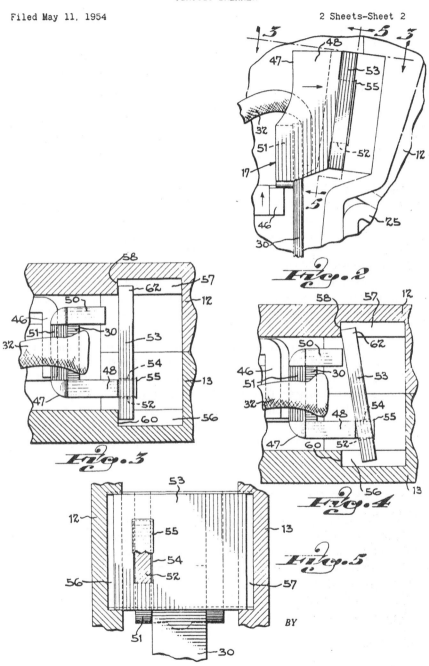

Figure 4.7 An improved electromagnetic tripping unit for thin circuit breaker (U.S. Patent Office).

and a similar need for relative motion to observe a gravitational field. Thus, Einstein noted, for an observer in free fall from the roof of a building there exists in his immediate vicinity, during his fall, no gravitational field. If the observer releases an object (neglecting air friction) it will remain at rest with respect to him and he with respect to it. The idea of a body being in motion and at rest at the same time are the antithetical concepts which Einstein used in his development of relativity. This was described by Einstein later as the happiest thought of his life.

The connection between two planes of thought that involves a seeming contradiction requires an especially high level of creativity. Black's development of negative feedback shows elements of Janusian thinking. The concept of reducing the output of an amplifier by introducing a fraction of it with negative phase relationship into the input is easily understood; but the reduction in distortion as a consequence is contrary to intuition. The improvement of the circuit breaker just described also involves this way of thinking. The solution occurred when I visualized the light-weight, hinged armature being pulled to the latch, being abruptly stopped, and the energy stored in that moving piece being delivered through the connection of the hinge to the member that was pulling it. During that moment, the two parts reversed roles and the trigger mechanism escaped.

In the referenced paper [19] the statement is made that Janusian thinking is not bisociation. It is described as a logical postulating of what on the surface seems illogical. Note however, that Einstein used a principle of the electromagnetic field to establish a principle of the gravitational field. He had found one connection between the two phenomena (or their associated planes of thought) and spent most of his life in an attempt to establish more. Janusian thinking is not bisociation but can lead to it.

BLOCKS TO CREATIVITY

Notice that in Figure 4.1 the looping mental activity on the vertical plane indicates a temporary hesitation to the bisociation. Such temporary blocks are normal. However, under some circumstances such blocks can permanently prevent the necessary connection between the two planes of thought. How do such blocks arise? Kubie [20] argues that there is no single cause but all can be lumped under the term *neurotic*. He cites examples of persons whose research went awry because of deep-seated emotional problems which caused prejudice, compulsion to spend mental energy on criticizing associates or proving a preconceived notion. On the other hand, a person who is at peace with himself because he understands whatever conflicts exist and can put them aside is free of this unwanted mental burden. The brain is free to act as a communication center processing bits of information on what Kubie calls the conscious, preconscious, and unconscious levels. On the conscious level a person deals with a subject in terms of communicable literal ideas and realities. On the preconscious level, he processes data at an extraordinarily rapid rate and with great freedom, assembling and disassembling many diverse patterns. On the unconscious level, without realizing it, a person uses his special competence and knowledge to express those needs indicated by his innermost concerns and his emotions. To the extent to which unconscious processes dominate the mental activity, the effective use of his preconscious thought process will be channeled to those problems. Not only are the products of preconscious thought vulnerable to distortions from the unconscious levels, the stream of activity itself must be protected from the same influences because creativity depends upon its free flow. The preconscious processes operate best when they are not restricted by the conscious and do not suffer interference from the unconscious. Perhaps you have experienced the technique of "sleeping on" a problem of deep concern with the happy result that the solution was obvious as you awoke the next morning. The activity of the preconscious does not depend upon our being alert or even awake.

There is another research report that gives insight to a cause of mental blocks. Hyman and Anderson [21] report tests whereby colored slides of familiar objects, such as a fire hydrant, were projected upon a screen and subjects tried to identify the object while the picture was out of focus. Gradually the focus was improved in discrete stages. The striking finding is this: If an individual wrongly identified an object while it was far out of focus, it had to be brought to a significantly better state of focus for him to correctly identify it than for others who had made no appraisal at all. A general statement would be that it takes more evidence to overcome an incorrect hypothesis than to establish a correct one. Or in words easier to remember, a false start can produce a mental block.

This discussion was included here to provide a positive basis for advice on improving creativity, not to worry you with the thought that inventing is difficult. First, since the preconscious is directed by your emotions, you must really want to invent to do so. Second, you must recognize those concerns that may redirect your preconscious activities—even against your will—and learn to set them aside. Third, learn to study the problem but do not decide too quickly on the mode of attack. More will be said about this later.

One of the most obvious blocks to creativity is caused by our education. This occurs because we become prejudiced that our particular area of engineering, the things we are expert in, is somehow the best. Electrical engineers look for the elegant solution only in terms of electrical devices or phenomena, mechanical engineers look to mechanical devices, and so forth.

A personal experience gives a good example of this prejudice. One of my patents involves a toy that was developed years ago incidental to Cub Scout activity. This is shown in Figure 4.8. It is a teeter-totter made of a bar magnet (27) which is positioned over a coil of wire (17). The pivots of the magnet (26) are slightly above the center of gravity of the rotating member. The coil is connected to a D-size battery (19) through a momentary contact switch (contacts are 22 and 23). When the switch lever (20) is pressed, the teeter-totter will rotate a bit but one closing of the switch does not accomplish much movement. Successive closing done in rhythm with the teeter-totter motion can increase the kinetic energy enough to completely turn the teeter-totter and its two occupants through a full revolution. The operation takes the same sense of timing that a child needs to "pump" a swing. The difference is that the toy requires only hand movement, not body movement.

After a toy manufacturer expressed interest in the device the decision was made to apply for a patent. During the initial meeting, the patent attorney asked if a similar play action could be gotten by a mechanical toy. I hadn't even considered that but immediately responded that it could not and even if it could it would not be as much fun.

I refused the first offer I received and in time learned how difficult it is to sell a toy. Toy manufacturers did not feel comfortable with coils, magnets, and low-friction pivots. On the other hand, plastic parts are their stock and trade. Several years after my patent was issued a toy appeared which could have been derived by analogs from mine (I am not implying that it was). This used a plastic bean pot mounted on pivots with center of gravity slightly below the pivots. A measure of plastic beans was provided to pile on the flat top of the pot. Each bean so added raised the center of gravity of the pot and bean system until the potential energy was sufficient to overcome pivot friction, rotate the pot, and "spill the beans." That is what the toy was called. It enjoyed a number of successful years.

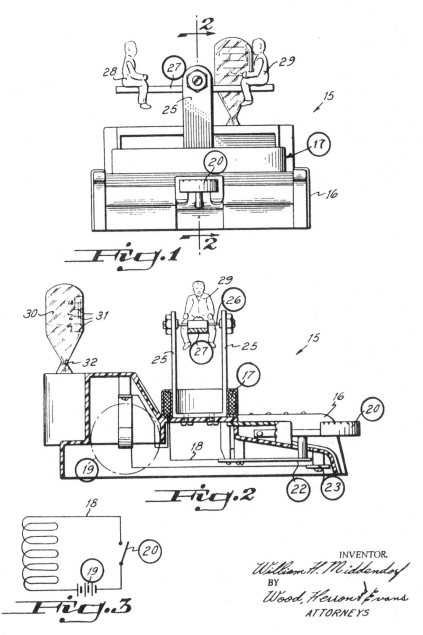

Figure 4.8   An electromagnetic game (U.S. Patent Office).

CHARACTERISTICS OF CREATIVE PEOPLE

As in all areas involving human endeavor, there is no complete agreement by investigators concerning the characteristics of creative people. Fortunately for our purposes, it is only necessary to consider the less controversial aspects.

According to D. W. MacKinnon [15] the creative individual enjoys esthetic impressions; has high aspirations; values independence and autonomy; is productive; has a high intellectual capacity; genuinely values intellectual matters; is concerned with his own adequacy; is dependable and responsible; has a wide range of interests; is ethically consistent; appears socially at ease; enjoys sensuous experiences; is critical, skeptical, not easily impressed; is candid in dealing with others; is talkative; and is generally introverted especially when engaged in creative activities.

Frank Barron [22] gives the following description of a creative scientist:

1. High ego strength and emotional stability
2. A strong need for independence and autonomy, self-sufficiency, self-direction
3. A high degree of control of impulse
4. Superior general intelligence
5. A liking for abstract thinking and a drive toward comprehensiveness and elegance in explanation
6. High personal dominance and forcefulness of opinion, but a dislike of personally toned controversy
7. Rejection of conformity in thinking (although not necessarily in social behavior)
8. A somewhat distant or detached attitude in interpersonal relations, though not without sensitivity or insight; a preference for dealing with things or abstractions rather than with people
9. A special interest in the kind of "wagering" which involves pitting oneself against the unknown, so long as one's own effect can be the deciding factor
10. A liking for order, method, exactness, together with an excited interest in the challenge presented by contradictions, exceptions, and apparent disorder

To these summaries, I would add that creative persons usually have ability and willingness to explore tenuous connections between only remotely connectable things. While the vast majority of such attempted connections lead to nothing useful, occasionally one yields a novel and useful insight into the problem at hand.

A very important characteristic for creative persons dealing with shape, composition, or physical interaction is the ability to visualize constructions in the medium in which they work. For example, the skilled artist must "see" the final picture before it is produced so as to evaluate the effect of every brush stroke. Earlier in this chapter Mozart was quoted as saying that "it stands complete in my mind so that I can survey it like a fine picture. . . . " Similarly, the interior decorator must visualize the results of choosing a certain sofa with certain rugs and drapes placed in a certain room. No one could earn a living as a decorator if he needed to see the choices in place before being able to decide whether or not the room would have the desired appearance.

Some designers work with systems or devices that require ability very nearly that required of a composer. They must be able to visualize relationships among ways of displaying information rather than things. The electronic circuit designer belongs to this group. A clear mental picture of component characteristics is required so that the effects of choosing one transistor over another or one circuit connection over another is readily apparent. Other designers work with systems or devices that require ability more like that of the artist or interior decorator, i.e., they must evaluate the effects of spatial relationship among materials of various shapes and of forces or potentials. Likewise, proper mounting of electronic components to provide adequate ventilation requires the ability to visualize spatial relationships. Thus, there are variations in the requirements to visualize, depending upon the job to be done.

A creative person does not hesitate to think unconventionally. On the other hand, a truly creative person does not select the unusual just because it is different. It must also be elegant. It is relatively easy to invent new devices or systems if being unusual is all that is required. For many years a popular satirical cartoonist with an engineering education named Rube Goldberg drew ludicrously complicated systems that accomplished useless or trivial results. There was creativity in the humor of the cartoons but certainly not in the invention of the systems. In fact, it was particularly unflattering in the 1930s and 1940s to call a design "a Rube Goldberg."

A creative person has a tendency to be dissatisfied with the products within his field. This is a natural consequence of being creative. So many alternatives are evident to the creative person that some other design flashes to mind as being more desirable. This characteristic is important because it may give you some insight into your potential as an inventor. Have you ever remade a new device, improvised to fix something that was broken, or simply made do with what was at hand to accomplish a certain task?

One of my students offered a good example of this. He had acquired a poster and wanted to hang it in his room without using masking tape. Surveying what he had available he used a safety pin as shown in Figure 4.9. He was careful to penetrate only half the thickness of the cardboard backing. The safety pin hanger worked as well as anything he could have purchased.

Last, a creative person maintains enthusiasm about his work, often in the face of disappointment. Creating something new requires full involvement of the skills of the inventor. Half-hearted participation will likely produce nothing of value. Those who lose interest quickly simply do not last long enough to invent. Furthermore, the inventor usually takes great pride in his own accomplishments. If a person does not really care whether or not he produces, the effort necessary to produce will not be maintained. Pride in accomplishment is a vital motivation.

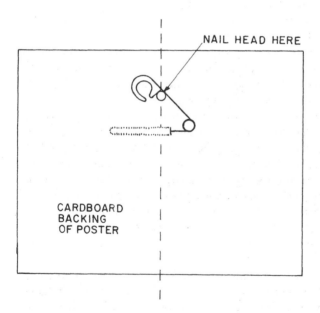

Figure 4.9   A student's improvised poster hanger.

CREATIVITY AND AGE

The effects of age on creativity was addressed in a book by H. C. Lehman entitled *Age and Achievement* [23]. In this study Lehman tabulated the number of creations within five-year intervals by his subjects, calculated the average number of creative contributions in each age bracket, and plotted the average number of contributions against the age brackets. His study covered many fields, such as chemistry, mathematics, fiction writing, and so forth. The interesting finding was that there is a certain range of ages, extending from the late 20s to the early 40s, which seems to hold as the period during which creative persons make the maximum number of contributions. Each field has its own particular range, which is shorter than the composite.

In the category of practical inventions Lehman shows that, based on 554 contributions by 402 inventors who were deceased at the date of his publication, the mid-30s are the most likely years for invention. The frequency of occurrence fell off almost as rapidly between the ages of 40 to 50 as it had risen between 20 and 30. Nonetheless, the data included some inventors younger than 20 years of age and some nearing 80. When comparing the most productive years of inventors born prior to 1750 to those born between 1830 and 1850, his data show that both groups enjoyed maximum output in their mid-30s but the earlier group remained much more productive between 35 and 65.

As to quality of contributions, Lehman shows that the 40 greatest inventions of modern (1953) times by 35 inventors occurred most often when the inventor was 32 years of age. The frequency of occurrence falls off even more sharply at less or greater age for this select group than it does for the larger group of "practical" inventions.

Lehman's findings have been substantiated by Bromley [24], who tested 32 men and 32 women in each of four age groups and graded ideas they generated as common or unusual. The youngest group with average age 27 had the largest number of ideas and the largest number of unusual ideas. The total number decreased somewhat with age but the most significant decrease was in the number of unusual ideas, especially in the oldest group whose average age was 72. The number of unusual ideas contributed by that group was less than a third of the number contributed by the youngest group.

Lehman [23] cites 16 possible causes for the decrease in contributions with age. However, these do not indicate a decrease in ability to be creative with age. Rather, a redirection of interest and effort seems to be the major factor. For example, he lists preoccupation with the larger affairs of successful men, less drive, decline in health and vigor, and neglect in staying abreast of the ever expanding state-of-the-art. Staying abreast of the state-of-the-art is, of course, more difficult as the change in technology accelerates. This was true in the comparison of inventors born before 1750 and those born before 1850. The ever changing technology is even more evident today.

These reasons given by Lehman for the decrease of invention with age is not subtle lessening of the ability to create but merely the obvious changes in the strength and available time one would expect to occur with age and success. Further, it must be remembered that those findings are statistical and cannot be applied to any one individual to predict his or her limits of creativity. This point is well emphasized by a list Lehman includes of 92 well-known older persons who made very great contributions when more than 70 years of age. Also, it has been noted that an effective stimulus for continued contribution is for the creative individual to deliberately change his field of endeavor.

INDIVIDUAL VS. GROUP EFFORT

For many tasks, team effort is accepted as being much more productive than the sum of the effort of the individuals. Does this hold for creativity? The theory advanced by Koestler [16] shows the advantage of group involvement as well as the major reason the group may not be successful.

As to the advantage, there is no doubt that the varied education and experiences of a group will increase the probability that the appropriate combination of ideas are stored in the minds of the participants. The problem is that the search for these tenuous connections of seemingly unrelated things cannot take place between two minds. They must be contained in one. There is a possibility, of course, that in exchanging ideas someone will describe just what another needs to establish an appropriate connection and, with further conversation, the group will arrive at the point of invention. There are indeed valid multiple-inventor inventions. However, they are probably rarer than the patent listings would indicate. Names are often included to avoid conflict or to reward those who brought the invention to a successfully engineered product whether or not they were true participants in the invention.

It is easy to speculate that Gutenberg had acquaintances who were familiar with wine making and had discussed the need for improving the printing process with them. However, his description was insufficient to direct the thoughts of the acquaintance to the force of the wine press and the description of wine making by his friend was not vivid enough to make the connection obvious to Gutenberg. Then suddenly, when Gutenberg saw the process, the similarity leaped to his consciousness.

There are other reasons that groups do not perform to their full potential in view of the extensive availability of knowledge and experience. This has to do with the tendency to be less interested in full dedication of oneself when a member of a group than when acting alone. An individual will spend extra hours thinking about the problem, work enthusiastically on it, and plumb the depths of the preconscious to find a solution. Members of a group are more likely to do what is necessary but not much more.

Data collected by psychologists [25] show that groups get more solutions to problems than do individuals but not more per member, and dividing the problem among individuals and adding all the answers gives an even higher total. Groups tend to correct each others mistakes, so the group judgment reduces quantity but improves the average quality. Individuals produce more good designs and also more bad designs. However, the truly bad designs are usually recognized as such and discarded. It is better to base your company's products on a small number of truly good designs than on a larger number of good but lower quality designs produced by group activity.

5

A Survey of Inventors

CAN INVENTORS DESCRIBE THEIR METHODS?

It may seem that the question of whether or not creativity can be taught and whether or not structured procedures stimulate invention should have been settled with certainty long ago. There are several reasons that this area of human activity is especially difficult to probe. One is that the inventor would probably prefer to think that his was indeed a flash of genius. Persons who consider themselves highly skilled in any activity eschew detailed explanations of their skill. A touch of mystery adds honor.

There is also a pragmatic reason for an inventor to avoid explaining during the life of a patent how the invention was made. Any explanation that makes the result sound like a logical conclusion of a sequence of steps reduces the results to engineering practice and gives grounds for an adversary to challenge the patent in court as being invalid.

As a class project each student in my product design course was required to identify an inventor by library research or through his cooperative job experience and to ask that inventor to complete a questionnaire. For reasons given above, the inventor and the patent he described as his most creative effort are not specifically identified.

The respondents to the survey represent a wide variety of backgrounds and present employment. There are outstanding college faculty, engineers and scientists from leading industrial research laboratories, others from small independent research and development laboratories, freelance inventors and engineers from typical midsize product-manufacturing organizations. In the matter of education, the respondents range from those with doctorates to one very prolific inventor who identified himself as a high school dropout.

THE RESPONSES

The information given here consists of

1. Number of patents on which the respondent is listed as sole inventor/same, but as coinventor
2. Time between recognition of need and occurrence of idea used to fill it
3. Technology area of the invention which the respondent believes to be his most creative effort
4. Method (if any) respondent uses to stimulate invention, or advice he would give student to increase likelihood of inventing new products or improving existing products

The information is given in the above order.

1. 4/2; six months; connector system for flexible printed circuit boards.
   Advice: Try to get ideas of what it takes to fulfill particular need even if not producible. Then, after you feel comfortable that your idea would work if you could make it, concentrate on how you can substitute manufacturable items for those which are difficult or impossible to make.
2. 1/0; one year; dermatology.
   Advice: Know the needs
   Interact with bright, creative people
   Work hard
   Continually have your ideas circulate in your brain
   Have good interpersonal relationships (especially at home)
   Take risks
   Go in different directions from majority
   Use resources available to you
3. Several/several; 20 years; acoustics.
   Advice: Rely upon "hands-on" experimentation
4. 60/20; one year; electron ballistics.
   Advice: Let yourself be puzzled about a problem that really fascinates you; worry about it off and on over a long period of time and try to get practical experience with it

5. 56/52; several days; data storage using magnetic domains.

   Advice: Respondent said his observation is that ability to invent is an inherited trait

6. 12/16; one year; keying system for musical instrument.

   Advice: Read about work of successful innovators, such as Edison, and accounts of great engineering projects, such as the Manhattan Project.

7. 1/0; two months; data coding for security.

   Advice: Brainstorming* is common practice here. However, the overriding factor is the selection of a working environment conducive to inventing. Ideally the company's product area should involve rapidly expanding product needs and the fulfillment of these needs should closely mesh with expanding new technologies. An example would be the application of microprocessors (new technology) to automobile pollution control (expanding need).

8. 5/14; six months; data storage.

   Advice: Be aware of basic needs

   Do not rely upon literature or others to find solution

   Look broadly at all technologies

   Develop a working knowledge of basic sciences or technologies related to your area of work.

9. 24/39; few weeks; rod drive for nuclear reactor.

   Advice: The company uses brainstorming sessions, but the best stimulation is to work in a location to which challenging problems and needs are brought. There, work closely with people who have a record of generating inventions. Also, obtain and study patents in your field of interest so that you learn how to present ideas.

10. 0/1; few weeks; pressure transducer.

    Advice: Creativity is strongly influenced by the intellectual stimulation and encouragement of the employer. Often the day-to-day activities leave little time for any attention to improvements.

11. Approx. 50/Approx. 50; one year; thermostat control.

    Advice: Recognize need and use by industry of a product. Determine if it can be manufactured at lower cost or improved to give prospective manufacturer an advantage over com-

---

*See Chapter 8 for description of techniques.

petitors. Select that product for intense effort. For example, could the residential door bell be redesigned so that it could be manufactured solely by automatic machinery? This leads to the requirement of a radically new way of producing sound.

12. 1/10; three months; automatic analysis of phosphor.

    Advice: List factors you wish to include in product, check literature for existing technology, discuss the problem with others in your field.

13. 1/1; three years; residential drain.

    Advice: Do not believe that any idea is too small to pursue or any technology too old to need improvements.

14. 12/5; two years; automobile parking time/cost.

    Advice: Must have desire and continually be on lookout for ideas used in other fields or products for possible modification or as stimulation to solve problem at hand.

15. 58/0; six months; steam-driven automobile.

    Advice: Learn much more about all materials and method of manufacturing. Learn to think without referring at all to the present state of any art. Consider the true basic need of the device or problem, list all possible avenues of approach, then analyze each possibility and proceed with the most probable solution.

16. Many/many; need does not always precede appearance of idea; drugs.

    Advice: Become familiar with prior art of your problem. Determine the shortcoming of a device or system. Make certain that analysis and reasoning are correct. Discuss your idea with a confidante who is capable of constructive criticism. Make certain your idea is practical and worth your effort; then try it, but do not be discouraged if it fails.

17. 21/2; discovery preceded need; piezo-optical cell.

    Advice: Brainstorming and involvement are used. However, as a general approach, define the need, the goal, the requirements, and the limitations. Concentrate on all aspects of the problem and get to know it thoroughly. Work to acquire a good understanding of related facts and physical laws that may give a solution. Recognize that the best solution may be in any discipline or be interdisciplinary. Let your imagination run free at first; consider practical limitations later.

18. 26/2; immediate; liquid crystal display.

Advice: To obtain valuable background do not discard products that fail. First, find out why they failed. This can give you ideas for new products or teach other facts that you would never learn from books. Then, when developing a new product, do not make a literature search until after you have exhausted internal resources. Remember: in developing a solution to a problem, keep it simple.

SUMMARY

It should not surprise you that there are many diverse opinions presented by the respondents. However, except for the one respondent who believes that ability to invent is an inherited trait, there is a great amount of agreement in these responses. It is masked simply by the different ways it is expressed. For example, interacting with bright, creative people, reading about successful innovators, working closely with people who have a record of generating inventions, and selecting a working environment conducive to inventing are all ways to develop the desire to invent. In similar fashion, note how many suggest that you build your store of information. One says by experimentation, another by looking to all technologies, by studying patents, by discussing with others, and even by using discarded devices to learn how things work.

One need not agree with all points mentioned by the respondents. It may well be that in dealing with creativity there is no single set of guidelines appropriate for everyone. These responses will be accepted for what they are, namely, the opinions of successful inventors. They will be of considerable help in developing the next chapter on how to improve your ability to invent.

6

How to Improve Your Ability

Theories of creativity, psychological research, examples of inventions, and the survey were presented to serve as a basis for the answer to the most important question this book can ask, namely, what can you do to improve your ability to invent? The suggestions given here should not be looked upon as quick remedies to be started after you decide an invention is needed. Rather, they involve a long-term commitment, a way of life. Most also require considerable effort. Some are easily drawn from what has been presented. Others reflect personal experience and resulted from introspection as the answer to the above question was considered.

The suggestions offered here often make reference to an employer. It should be recognized that the advice is just as valid for the independent inventor if he views himself as both employer and employee.

SUGGESTIONS FOR THE WOULD-BE INVENTOR

1.   Develop a desire to invent. This desire can be enhanced in a way suggested by one of the survey respondents, namely, by reading about inventors and important engineering projects. Trade magazines and periodicals often have articles about inventors and inventions of note. Also, you have seen by references given here that business magazine articles and even newspapers give interesting accounts of the accomplishments of inventors. However, do not simply wait to run across such stories; seek them out. Another way to increase the desire to invent is to determine what it will mean to you in your job and leisure time situations. The subject of compensation will be discussed later,

but there is little doubt that patents can have a positive effect upon the progress of your career. The reason it is so important to develop the desire to be an inventor is that this will influence you to meet the challenge when a need is identified. If inventing means nothing to you, you will not do it. Furthermore, your desire must not be lip service. It must be strong enough to involve the rapid flow of preconscious mental activity described in Chapter 4.

2. Avoid becoming involved in emotional causes that will consume your mental energy. The person who is spent by frustration over political disagreements or crusades for a particular solution to social needs will have used his creative powers in those concerns. He will not have the time and energy to be creative in the engineering sense. However, very important words in the prohibition are "Do not be consumed." It is not meant to say that you should neglect family, civic responsibility, and charitable involvements. This would run counter to the advice of one of the survey respondents who stated that you should have good interpersonal relationships (especially at home), or as Mozart is quoted as saying in Chapter 4, "when I am . . . of good cheer . . . my ideas flow best and most abundantly. . . ." In fact, a well-balanced relationship with family, community, and society promotes the freedom of mind conducive to invention.

3. Become familiar with inventions by reviewing the *Patent Gazette* [5] regularly. Or, if your employer maintains a file of current patents, study them. Either activity will make you familiar with the state-of-the-art and the kinds of problems others in the industry are addressing. You will become more conscious of invention as an activity intimately involved with your work. The very fact that you become more aware of patents will increase your disposition to invent.

4. Do whatever you can to become familiar with unusual physical phenomena. Remember Carlson developed xerography from the earlier work done by Paul Selenyi on the use of electrical charge to produce images. Notice how often in this book examples of the most significant inventions include a description of the inventor happening upon a reference to physical phenomena later used successfully in an invention. A convenient listing of physical phenomena is given in a book by Alley and Hix [26]. However, trade journals and professional journals are other important sources. Often such publications report phenomena which the editors feel are not widely known. If a library is available to you, make it a habit of visiting it periodically with the mission of finding at least one physical fact previously unknown to you. Whether

or not a library is handy do not pass up an opportunity to purchase any book that describes unusual phenomena. Invention is made by applying a seemingly unrelated solution to a recognized need. Gutenberg's use of a wine press as a starting point for the printing press should make clear that the more information you have absorbed the more likely the tenuous association necessary for invention will take place.

5.   Establish a depth of understanding of all phenomena that comes only with true insight into their operation. For most inventors it is more important to be able to visualize how physical events occur than to be able to express them mathematically.

6.   Realize that there may be areas of engineering and science with which you should be familiar but which were not included in your formal training. In an effort to allow students to personalize their college program some schools allow so many electives that it is possible to graduate with a named degree but without having done course work in a significant part of that discipline or a closely related discipline. If this is true in your case, a remedial program—perhaps through evening college or self-study—is appropriate. Also, other areas of engineering or science may prove to be the source of vital information. Who would have predicted that a book on visible light lenses would have provided the basis for Dr. Winston Kock's [27] solution to the direction of communication signals? Yet, as described in Chapter 8, this was the case.

7.   In agreement with the advice of a survey respondent, never discard products that failed before finding out how they worked and why they failed. A worn-out automobile battery regulator, a toaster, or any such device should be taken apart and studied, as a medical student dissects a cadaver, to learn how every part of the body functions. If you were to study only one such device each month your wealth of knowledge would soon be superior to that of your co-workers. Charitable organizations which accept used material for resale and, of course, friends and relatives are good sources of worn-out devices.

8.   Improve your hands-on working ability. Jacob Rabinow, an inventor who is referred to several times in this book, casually mentions [28] that he annealed, reshaped, and then rehardened a shakeproof washer as he solved a difficult quality control problem associated with self-arming bombs. As described in Chapter 3, Black's and Carlson's activities included experimentation, model building, and so forth. This skill runs counter to the present day deemphasis of engineering

laboratory courses but is a valuable asset for the inventor. The excitement of a promising idea is strong motivation if it can be pursued while still fresh. If a simple task such as changing the shape of a washer presents the problem of scheduling it through the company's model shop or, for the independent inventor, of locating a vendor who will supply a small number to your specifications, the idea will probably be discarded without trial. In fact, it is possible that an elegant solution may be blocked out of your thought processes if it causes frustration.

Another important reason for a would-be inventor to become a hands-on experimenter is that valuable feedback is generated by carrying out the experiment that cannot be generated in any other way. The engineer who directs technicians to build and test the model will receive a report carefully stating that the invention did not work but offering a little insight as to what possibly would make it work. It is fairly common for a new device or process to fail, not because it is basically defective, but because some minor problems must be resolved.

9. Take appropriate steps to acquire the necessary facilities to try out your ideas. Tools, model building materials, test equipment, and measuring instruments are needed to pursue an idea whenever it occurs. These need not be research quality equipment or equipment convenient to use. Simple models can often test an idea sufficiently to determine whether or not you are on the right track. Later you will read about how Dr. Winston Kock used pearl beads painted with conductive paint and mounted on wood sticks to test out his invention of the electromagnetic lense that now transmits television signals across the country.

10. Immerse yourself in a climate of creativity. Work for a company where new ideas are welcomed and appreciated. It is difficult to become interested in pursuing the program of self-development being suggested here if there is a feeling that it will not result in respect, status, and appropriate compensation.

11. Recognize that opportunities vary depending upon the maturity of the technology. The newer field usually presents the better opportunity to do something novel. This is not to say that invention will come only to those who deal with the latest technology. The changing economic scene is also the basis for new needs to develop. For example, mechanical fastening is probably the oldest technology of human history. However, new fasteners using less material or requiring less

work for a specific application are constantly being developed and patented. Remember the example in Chapter 2 of the self-locking nut whenever you think that there is nothing left to invent in your industry.

12.   Realize that there are jobs in industry which are not conducive to invention. For example, an engineering manager should expect to direct his attention to budgetary matters, personnel problems, equipment needs, reports for upper management, and a host of other responsibilities. Certainly, a unique solution to a product- or process-related problem can occur to an individual so burdened. However, the probability is certainly less than for an engineer primarily involved with the company's product or the manufacture of it. The point of this advice is that an engineer may be attracted to a lower management job with its diversion from inventing too early in his career. A delay until significant engineering accomplishments can be made may give the credentials which will result in a much greater opportunity for management later, if so desired.

13.   As soon as you identify a need which is important enough for you to initiate invention effort, start a notebook or folder in which every idea is recorded. If the idea is discarded make an entry giving the reason. The reason for rejection may not be valid at some later date because of a change in technology or the economic situation; or putting two of your previous ideas together may give the elegant solution you seek. Another advantage is that simply reviewing such a file later will renew your enthusiasm if the need still exists. You can understand the need for such records if you remember the discussion of Carlson and Black. Both labored so long over their great inventions that it would have been impossible to remember in detail all that they had previously tried.

14.   Set aside a portion of each work day or work week—depending upon your individual situation—to consider how you can improve your company's products. Knowing in what way the products are less than ideal will make you a better engineer and may provide that opportunity for a significant improvement. Reread the story of Harold Black in Chapter 3. Imagine that young engineer addressing the problem of 3000 channels with 4000 miles of transmission while his superiors were content to consider three channels with 1000 miles of transmission. He clearly understood the problem of future telephone communication better than those who were supposed to direct him.

15.   Above all, take your time in seeking a solution to significant problems. Black found the key to feedback amplifiers after eight

years, Watt had worked on the steam engine for several years, and Carlson's patience seemed inexhaustable. The time frame, of course, must be appropriate to the importance of the need. However, judge wisely. The lock nut invented by Horace Holmes was not his first invention in that field. It involved years of seeking the best solution. And although it may seem rather simple, realize that billions of nuts are used annually. That is what makes it important.

SUMMARY

You probably feel overwhelmed by this list of activities, perhaps even a little angry that anyone would suggest so much. After all, Roberts did not do this to invent his easy-release socket wrench, no evidence was offered that Holmes is an expert on other than certain kinds of fastening means, and even Carlson seems to have made the great invention of xerography by sticking strictly to reproduction of images by electrical charge transfer.

The list of 15 suggestions offered above is meant to be extensive. I would like to be able to say that it is exhaustive, i.e., it guarantees invention. Of course, I cannot say that. However, it is everything that I have used or that I know others have used. I believe that anyone who follows all this advice will be able to invent. I also believe that many who follow only part of it will invent. And some, will invent who have done none of it, such as Roberts.

Look upon this advice as a continuing program of self-development. It is not meant to make you a fanatic about inventing or a person who believes his every act must be creative. There are occasions when the usual way *is* the elegant way. Prepare yourself for the several times during the remainder of your professional career where a truly significant opportunity arises to satisfy a need. It can be something as urgent as protecting life or as light-hearted as making children laugh. Both are important. Invent to make contributions to society. Do not invent simply to invent.

7

Choosing the Best Strategy

A WARNING

Almost as soon as you are aware of the need for an invention you can fall into an insidious trap with no clue that anything is amiss. This has to do with the strategy you use to fulfill the need. Often the way you receive the information which leads to identification of the problem makes the approach you select seem obvious and yet it will be the wrong approach. It will be wrong because the resulting invention will not be as elegant as would have been possible with another strategy. An example will help to clarify this important point.

Jacob Rabinow is one of our country's outstanding inventors. One of the inventions he mentions in his after-dinner talks will be repeated here.

In 1945 he received a waterproof watch as a gift from his wife. That was when good watches kept time by the periodic motion of a spring and flywheel. If they ran a bit fast or slow there was a lever on the back of the mechanism that could be moved one way or the other to change the period of oscillation. Of course, there was no way for the owner to tell how far to move the lever to compensate for a given rate of error. The correction, if accomplished, was done by trial and error.

The main task in making the adjustment was removing the back of the case. That required a special tool. Nonetheless, many people used whatever was handy and damaged the case as well as their composure. After several openings of the case, Rabinow had become aware of the need. But that is the point of the story. The obvious approach would have been to spend his inventive talents on an easy-to-open waterproof case. He had the good fortune of not pursuing that course. Instead, he selected the strategy to eliminate the need to open the case. He recognized that whenever a person corrects the setting of a watch, he introduces motion into the mechanism by rotating the stem and the direction of that motion is a measure of the error the watch had made. There were a few complications, such as setting the correct time as an area switches to and from daylight saving time. However, these too were solved.

The invention never sold well on watches for marketing reasons, which Rabinow explains with a great deal of humor. However, it is used on automobile clocks and changed these from a useless item to one that really does keep the correct time after a series of resettings. Every automobile clock now made contains this invention.

The reason this trap was called insidious is that an inventor may be well satisfied with himself if he makes a product improvement while following a lesser strategy. Remember the research by Hyman and Anderson reported in Chapter 4; a false start can produce a mental block. The inventor probably will not even stop to consider the possible existence of another strategy. Yet the advice which I want to give with more emphasis than anything else stated in this book is that you should stop very early after identifying a need and ask yourself how many strategies are available to you. Besides asking how many ways you can fulfill a need, always ask if the need can be eliminated.

This subject is so important that it has been given the treatment as a full chapter even if a very short one. What needs to be said has been said; now it must be remembered.

RABINOW'S PATENT

Figure 7.1 shows the first page of Rabinow's patent (2,542,430). Figure 1 is the back of the typical mechanical alarm clock. Part 2 is the stem used to set the clock hands to the correct time. Part 6 is the winding stem which Rabinow replaced by a two-position stem to wind the mainspring when pulled out and set the clock hands when pushed in. The arcuate slot near the bottom of Figure 1 shows the lever which adjusts the rate of oscillation of the flywheel. The operation of the invention can be understood from Figure 2 of the patent without worrying about details. Rabinow connected the time-setting stem to the clock hands by gear 3 and to the rate lever by a series of gears terminating with 39.

Claim 1 of the patent describes the invention as follows:

A timepiece comprising time indicating means, control means for setting same to a desired value, rate regulating means for increasing

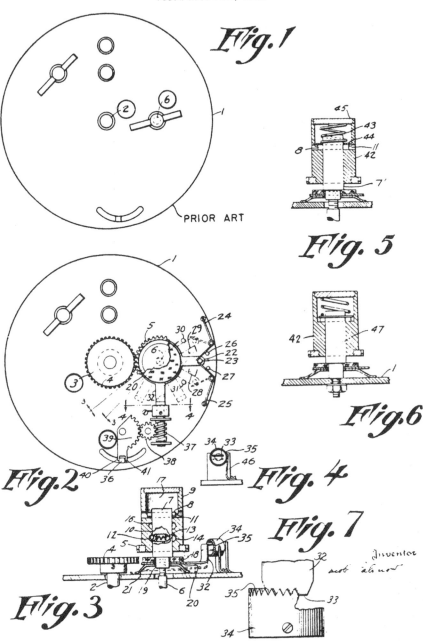

Fig. 1

PRIOR ART

Fig. 5

Fig. 6

Fig. 2

Fig. 4

Fig. 7

Fig. 3

Inventor

Figure 7.1   A self-correcting timepiece (U.S. Patent Office).

and decreasing the running rate of the timepiece, adjusting means operably connected between the above control means and rate regulating means to change the adjustment of the said rate regulating means whenever said control means are operated; said change in adjustment being in a direction dependent upon the direction of the difference between the original reading of the said time indicating means and the subsequent reading to which the time indicating means are set, said adjusting means including means arranged to cause the said changes in the adjustment of said regulating means to be of a predetermined fixed amount.

8

Methods to Stimulate Invention

The suggestion that a structured method can lead to invention is a subject that causes controversy when creativity is discussed. One of the most prolific of the respondents to the survey, a freelance inventor, wrote, "I have yet to see such a method produce anything of value," but then gave a few steps including "list all possible avenues of approach and analyze each." This is in fact what Carlson did as the first step in the development of xerography. Another respondent wrote that his "observation is that the ability to invent is an inherited trait." On the other hand, half of the respondents indicated that some structured way of proceeding is a valuable aid to invention. Some emphasized their suggestions as step-by-step techniques by numbering each step; however, others mentioned such standard techniques as brainstorming and involvement, which will be discussed. Perhaps the proper attitude regarding structured methods was expressed by a student recently after a lecture on the subject when he said, "If there are those so creative that they do not need procedures to follow, fine—let them do it their way. For others, a way to guide ones thoughts might be necessary."

Structured methods may help principally by guiding the inventor to those intersecting planes of thought that are associated with the plane of interest and thereby promote the bisociation. Also, it is probable that such activity will promote interest and complete mental involvement in the needed invention to the point where the preconscious activity will be directed to that end. With that in mind, the following are offered from a long list of possible procedures [29].

A GENERAL METHOD

First a procedure will be given which is useful in itself and can be readily changed to accommodate other methods. The steps are

1. Define the need.
2. Gather information.
3. Review the elements of the problem repeatedly.
4. Try for a construction that works even if it is not as elegant as you had hoped.
5. Try for an unusual and elegant construction even if it does not satisfy all specifications or work perfectly.
6. Repeat steps 1 to 5 several times.
7. Direct your attention away from the problem for some time—weeks if possible. This is time to let your preconscious mental activity take over.
8. Get back to your problem when you feel enthusiastic about working on it. This may be a signal that you are ready for some creative insight.

Note that this method is similar to one suggested by a survey respondent who advised that the inventor seek a solution which fulfilled the need—even if not producible—with adjustment later to get both a creative and practical solution.

ADAPTATION

A very important method which can be used in conjunction with the general procedure to develop the constructions called for in steps 4 and 5 is called adaptation. It is probably used in its explicit or subtle forms more than any other method. In this method a solution of a problem in one field is applied to a similar problem in another field. For example, a pistol grip is one of the most comfortable handles made for the human hand. It was, of course, originally developed for guns but is now found on soldering irons, electric drills, movie cameras, electric shavers, and even on ammeters. This is a trivial example of adaptation because the use of a certain type of handle solves the same problem no matter what is held, i.e., it was not necessary to discover the relationship between the functions to be performed. In most situations where adaptation is used the significant step is to recognize this functional relationship. Consider mounting a clip to a sheet of metal. By recognizing that a hole in the sheet of metal is comparable to the opening in a "loafer"-type shoe the connection between mounting the clip and retaining the loafer is established. The bisociation which follows is that the clip can be given the shape and resilience of a foot and retain the sheet of metal *to it* just as the foot retains the shoe (see Figure 8.1).

Gutenberg's use of the principles of the wine press to develop the printing press is a good example of adaptation. He experienced the operation of the wine press by chance. However, in a structured way he might have listed the characteristics which he could have identified as necessary for the printing press and methodically searched for devices that shared those characteristics. We assume that a thorough search would have led him to the wine press.

As another example, Dr. Winston Kock discusses in his book [27] how he developed microwave lenses to focus electromagnetic waves. These lenses are used to focus communication signals from one relay tower to another. A network of such transmission now extends across the country. The development had its birth, Dr. Kock writes, more than a decade before the need for such a lens surfaced when he read the book *Optik* (1933) by the German scientist Max Born. Born showed that the action of the optical lens can be explained by assuming the material (usually glass) is an assemblage of tiny, closely spaced, reradiating particles, each acted upon individually by the electromagnetic action of light waves. In terms

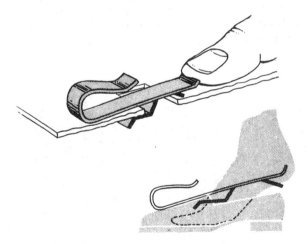

Figure 8.1 Escutcheon clip mounted on sheet metal. (From W. H. Middendorf, *Engineering Design,* Allyn and Bacon, Boston, 1969. Copyright W. H. Middendorf.)

of the wavelength of visible light the particles of solid glass are sparse, i.e., they are separated from one another. To the scale of microwave radiation a comparable lens could be made by small conducting spheres mounted on insulators. Dr. Kock's first embodiment was a large number of imitation pearl beads coated with conducting paint and mounted on thin sticks. This is shown in Figure 8.2. He writes that the focusing action from this first model was quite pronounced. Numerous improvements and several different kinds of lenses were patented by Bell Telephone Laboratories as Dr. Kock's research continued, but the basic contribution had been completed by recognition of the structure necessary to react with mocrowaves in the same way that visible light reacts within glass.

The adaptation method is important enough that it has spawned research in an area called bionics, which is the study of living things as engineering prototypes. In other words, humanity is actively seeking to learn the problems confronting the lower forms of life and the solutions which nature has evolved in response to these problems. As a result of this type of research, a ground speed and altitude indicator for aircraft was fashioned after a beetle's eye, and a heat sensor was based upon the heat-sensing organ of a Massasauga rattlesnake. As you repeatedly perform steps 4 and 5 of the general method to stimulate invention which was given previously, you should consider the adaptation method as a means of finding an idea upon which to base an invention. Adaptation does not mean merely copying an obvious solution. The creativity resides in recognizing a subtle connection which has escaped the notice of others.

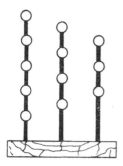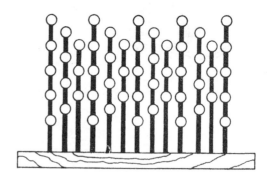

Figure 8.2 Conducting spheres arranged to bring microwaves into focus.

ANALOGS AND DUALS

A method that can be considered related to adaptation is one whereby a systematic change can be made in the variables of the defining differential equations to express the phenomena in the same discipline (dual) or in a different discipline (analog). These techniques are well known to engineers and will not be taught here. It should be sufficient to remind you of them and offer some comments. For example, a system of duals exist between electric networks whereby any variation of current in one network can be replicated by a variation of voltage in the dual network and vice versa. Furthermore, for planar networks there are very simple ways of finding the dual of any given network. The same statement of duality can be made regarding force and velocity in a mechanical system but there is no simple way to find the dual system.

Using analogs, an electric network can be found where current and voltage vary precisely as force and velocity, respectively, in a mechanical system and another network (the dual) where voltage and current vary as force and velocity, respectively. One advantage of using duals and analogs is that the performance of a system may be easily understood in one discipline whereas in the other it is not as obvious. Another advantage is that some phenomena are more easily produced than are the analog phenomena. An example will be given to clarify this technique.

In his talks after-dinner Jacob Rabinow tells of his invention of the magnetic clutch. This happened early in Rabinow's career while he was an employee of the National Bureau of Standards. Winslow had discovered the electrostatic particle clutch while working with clutches of various designs. Winslow found that if particles of starch or limestone were embedded in a film of oil between two plates and a high voltage was applied across the mixture, the particles chained up into a kind of conglomerate that made the two plates bind together. He applied for a patent on this device and after some delay it was issued. However, tests showed that the clutch was capable of transmitting only modest amounts of power. Finally, it reached one of the consulting firms to Rabinow's group at the National Bureau of Standards. On one occasion, while working with it in the laboratory, it suddenly occurred to Rabinow that if electrostatic attraction could be made to do the job, electromagnetic attraction should be much more intense. The permeability of iron is about 2000 whereas the permitivity of starch is about 2, so that for the same field stress, the magnetic clutch should be able to withstand a thousand times more sheer force. Besides, an intense magnetic field offers none of the safety or material degradation problems that attend a strong electric field. The invention was an instant commercial success. The magnetic clutch is easily controlled by electrical means, is efficient, does not wear out, and never needs adjustment. The compensation Rabinow received for the invention will be discussed later.

BRAINSTORMING

Does it help to interact with others regarding the need which you identified? In fact, it helps in several ways. First, vocalizing the problem helps get it more clearly set in your own mind. It automatically increases your involvement with it. Second, another person can bring a fresh point of view to the problem. Different background, education, and experience may enable him to give valuable direction toward a solution which would probably not occur otherwise.

About 1939 a method called brainstorming [30] was formalized. Later in the 1950s it received such media attention that it was thought to be the answer to all our creativity needs. The procedure to initiate brainstorming is simple: gather a group of people with diverse backgrounds, allow no critical comments, and encourage ideas to flow freely. Later the ideas are evaluated by the leader or a committee to judge their potential for success. That step in the procedure may impose an important limitation on the method. The talents and open mindedness of the judge may be more important to the outcome than the quality and quantity of the ideas. Nonetheless, the technique is worthwhile and is mentioned more often by the survey respondents than any other technique, although some respondents cautioned that brainstorming as used in their companies may be different from the generally accepted procedure.

Brainstorming holds most promise for the simple problems which do not require in-depth analysis or careful evaluation of successive steps. For example, "how shall the cover be fastened to the widget for least cost, durability, and ease of maintenance?" is well suited to brainstorming.

For more involved problems, brainstorming can work if a small group can be assembled whose members have learned to give freely of themselves for the common good or the progress of the company. However, in general the disadvantages expressed in Chapter 4 regarding group vs. individual effort apply to reduce the effectiveness of this technique. It is my opinion that, for solution of engineering design problems or inventing, brainstorming is most useful as a means to exchange information and can lead the principal person involved, the inventor, to the realization that he should study some of the phenomena identified during the brainstorming session as possibly being important to the solution of the problem.

SYNECTICS

A method that combines brainstorming and the use of analogs was developed by William J. J. Gordon [31] in the late 1950s. This was the result of study of the creative process by replaying tape-recorded sessions during which novel ideas were generated. Several of Gordon's findings [32] confirm points which have already been made here. One has to do with the use of analogs. Gordon found that in virtually every case he studied an analogy had been the key insight that led to the discovery. A second point confirmed by his research was the widespread occurrence of paradox (Janusian) in problems requiring a creative solution. Finally, in the structured method that Gordon developed note that each participant suggests an "ideal" solution to the problem even though it may be totally impractical. This is also a step in the general method already given except that the suggestion there is meant to be as practical as you can conceive within the bounds of novelty and functionality.

Gordon's method starts with the leader describing the problem in terms of the underlying principle involved. For example, parking cars might initially be described as storing objects. The steps of the synectic technique are:

1. The leader defines the problem.
2. The group explores preliminary ideas by the participants who are not expert in the problem area. The critique of the expert is invited. This is called the *purge*. If a valuable suggestion is made, it is written down for later consideration.
3. Each participant is invited to formulate and define an idealized solution to the problem.
4. The leader selects one of the idealized solutions and asks for a transformation into a world he specifies. After a number of ideas are expressed the leader may call for one of these to be transformed into still another world.
5. After the group explores analogous problems through one or more transformations the leader asks how the ideas are related to the problem at hand. This is the force fit. If the group is unable to suggest a solution, the leader may return to step 4 to select another idealized solution or the entire cycle may be repeated with a new leader.

A demonstration of the method was given on a sound record included in the March 1968 issue of the *Journal of Engineering Education.* The problem was to develop a method of sensing weak areas in cloth as it is woven. The analogy used was personal feelings and this led to the idea of the cloth feeling sad about its defect which in turn led to an expression of sadness when President Kennedy's body was returned to Washington, D.C. on Air Force I. A play on words then led to forcing air through the cloth. Any weak part due to inadequate weave would allow increased passage of air which could then be detected during manufacture and immediately corrected. It is not known whether this was an actual successful session or contrived. The mental gymnastics required to go from defective cloth to Kennedy's death to an airplane named Air Force I to solution of the problem seems less likely to occur than simple consideration of ways to detect imperfections.

The synectic method has been described as vague and a technique that seems to work best when Gordon is personally directing the session [32].

INVOLVEMENT

Involvement is most useful with mechanical devices. The method assumes you will follow the general method given earlier and aids in carrying out steps 4 and 5. The central idea is that you must visualize yourself as being part of the mechanism. For example, suppose the need is to develop a way to remove dust and other particles from a phonograph record while it is playing. If you think of yourself as a little person cleaning the groove ahead of the needle you may be able to visualize how you would blow the foreign particles from the needle's path, and progress from that idea to a jet of air from a nozzle attached to the phonograph arm adjacent to the needle. Or, if you remember how you clean the basement or driveway (there is a hint of adaptation here) with a water hose, you might consider using a fine jet of water to move the dust and dirt from the record groove. The water and foreign material would then be recovered by a suction immediately ahead of the needle.

AREA THINKING

Area thinking is a technique which was taught by Professor John Arnold in the 1950s. He was a staunch advocate of the proposition that creativity can be taught [33]. The objective of this approach is to improve an existing product by concentrating on one area at a time which is important to the consumer. The areas to be considered include cost, performance, function, appearance, safety, repairability, and many others. The point is that, as you try to find the basis for an improvement, a conscious reminder that these areas of thought probably do intersect with that of the product you are considering may result in a bisociation. To look at this another way, by identifying lower cost, greater safety, or any other useful characteristic as a probable need of a product the first difficult step of the invention process is tentatively accomplished. This shifts the burden of creativity completely to satisfying that assumed need in such a dramatic fashion that its existence is thereby proved. Of course, often the result is verification that the product was already optimized.

One caveat is that the would-be inventor should not run down this list as if he were reading items of laundry. It may be well to consider ways to reduce cost over a period of, say, two months and then consider safety for a like time. As has been said earlier, take your time. Very few inventions happen on the spur of a moment.

FUNCTIONAL SYNTHESIS

The last method to be discussed was called "orderly creative inventing" in a 1957 publication [34]. It was identified by introspection after my first invention (the one on the circuit breaker magnetic circuit described in Chapter 4) was made. Discussions with G. T. Brown of National Cash Register Co. of Dayton, Ohio, who had similar ideas, led to the development of the method.

For the most part, it should be applied by individual effort because it relies upon precise, careful reasoning rather than a rapid fire vocalization of ideas. The most important part of this procedure is the act of describing a sought after device in terms of functional requirements. Descriptions of presently used methods of fulfilling needs should not be used in this step.

Engineers are taught to think in terms of needed functions in product or process design. Thus, this method is not a great departure from the usual mode of operation. In fact, the method is probably used by many engineer-inventors who do not realize that it is a structured method. Gutenberg's quick response to the action of the wine press indicates that he had clearly defined, in his own mind, the functional needs of the machine which would print.

The advantage of working only with the identifiable required functions is that all the embodments used in the past to provide those functions are stripped away and the inventor then concentrates on all other possible ways of providing them. This enables an engineer to start with whatever is presently used to accomplish the task, strip it of its present structure, and reconstitute it in a different form. There is no doubt that this invention procedure will result in novel devices. The characteristics that cannot be guaranteed are that the embodiments suggested by it will be within the state-of-the-art, economic, safe, convenient to use, and so forth. These characteristics are not necessary to have an invention but they are necessary for it to be economically successful.

The method consists of six steps. The first are identical to the general method discussed previously. The steps are

1. Define the need.
2. Gather information.
3. Divide the system into subunits.
4. Describe each subunit by a complete list of its functional require-
    ments.
5. List all the ways the functional requirements of each subunit can be
    realized. Each is a partial solution.
6. Study all combinations of partial solutions.

The third step, dividing the system into subunits, is necessary. In any device or system there is a key part or subunit whose characteristics influence the other parts to a major extent. This should be separated from the rest and treated first. Then proceed to another subunit. For example, in developing a new automobile, the engine is one subunit, the passenger compartment another, the coupling between earth and vehicle (usually wheels) the third. Which of these subunits is most important in influencing the overall design? Also, Figure 4.5 shows the subunits of a circuit breaker. Any product, big or small, simple or complicated, can be subdivided in this way.

The fourth step, describing the subunits by functional requirements, is the all-important step that must be done carefully. It is essential that each subunit is described by what it must accomplish and not in terms of a possible realization. If you do this skillfully, you will free yourself from thinking in terms of how the need was satisfied in the past. For example, the wheels of an automobile have the function of providing low-friction support of the vehicle for relative motion with the earth. They need not be wheels at all. How else could this be done? How about using legs, or a pad of air? Do you see how you start thinking of other ways to accomplish the same results when you describe the subunits by their functions?

The fifth step is simply to list, as briefly as possible, all the ways in which you can satisfy the functional requirements of step 4. This was partially done for the automobile when legs or an air pad was suggested. The longer the list of alternatives, the more chance for success. Here is where brainstorming can be most useful.

In the sixth step consider all combinations of the ways which were listed to realize each subunit. This results in many solutions. Some will most likely be old solutions, some will be ridiculous, but hopefully at least one will be novel, useful, nonobvious, and, of course, physically realizable and economical. These are the main requirements of a successful invention.

As an example, consider the invention of a new can opener. Definition of the need is step 1. Now by selecting the simplest can opener (see Figure 8.3) available the necessary parts (the subunits) as well as the functions of each part become apparent.

For the opener shown there are only two essential parts. One is solely to separate metal but you must recognize that the function of the handle is at least twofold. It enables the user to position the opener and it enables the user to apply power.

Step 5 is the search for all ways to provide these functions and is shown by Table 8.1.

The listing of Table 8.1 is not exhaustive. With a little thought you can

Figure 8.3   A Simple can opener. (From W. H. Middendorf, *Engineering Design*, Allyn and Bacon, Boston, 1969. Copyright W. H. Middendorf.)

Table 8.1 Search for All Ways to Provide Basic Function of a Can Opener

Part	Function	Realization
Subunit 1	Separate metal	1. By shearing 2. By tearing 3. By fatigue 4. By melting 5. By drawing thin 6. By chemically eroding
Subunit 2	Apply power	1. By hand 2. By electric motor 3. By hot wire 4. By hydraulic motor 5. By flame 6. By chemical reaction 7. By mechanical vibration 8. By laser
	Position	1. Bring can to opener 2. Bring opener to can 3. Have opener built on can

Source: From W. H. Middendorf, *Engineering Design,* Allyn and Bacon, Boston, 1969. Copyright W. H. Middendorf.

certainly add more ways to separate metal and more ways to apply power. Separate brainstorming sessions on how to realize each function would add many alternatives. In step 6, evaluate all combinations. For example, combination 2-1-3 in Table 8.1 describes an opener which tears metal by manual power with the opener built in place on the can. Beverage cans are now made in this way. Many other combinations of solutions will describe openers already in use. Others will describe openers that are impossible or impractical. However, there is a possibility of some combinations of alternatives that are practical. Furthermore, if they are not practical today they may be in the near future as the exploding technology gives us many improved techniques. For example, you may have been thinking of the flame from a match when you considered item 5 under subunit 2 listed in the table. Suppose, however, it would be possible to have a flame a few thousandths of an inch in diameter, a quarter of an inch long, and 3000 K at its tip. In that case 4-5-1 would be a possible combination. Perhaps an inexpensive laser will someday be used in this appliance.

It is important to note that, although this method is described by steps, you should not assume that once a step is given some attention you are necessarily finished with it. For example, it may now be appropriate to go back to step 2 (gather information) and find out more about how to produce very small intense flames.

SUMMARY

The methods which can be found in the literature as aids to stimulate invention make a long list indeed. It would take a lifetime to give all that have been published a fair trial. Furthermore, I have noticed that the person most vocal about the value of a particular method is the inventor of that method. As stated earlier, Gordon's method reportedly works best when Gordon is in charge. Others report it as vague. I have found functional synthesis to be clearly superior to all others. When the method was published there were approximately 200 requests for reprints. However, I have seen only two references in other books on creativity. Rabinow [3, p. 26] said he has tried to find out what a "morphological" approach is and has never gotten a straight answer, yet Zwicky [35] claims to have invented the method and has made profound discoveries and inventions with it. And so it goes!

From all of this I have come to the conclusion that the most effective way for each inventor to stimulate invention is to develop his own method. The ingredient that must be present is a consuming, sincere desire to invent. Rabinow [3, p. 90], echoing the advice of King Arthur in Camelot, says to "simply love them." Benjamin Horvay of General Electric says, "Force yourself to generate at least seven alternatives . . . they probably will not lead directly to a solution but this will get you so immersed in the problem that the subconscious [Kubie's preconscious] mind will take over and solutions start to emerge into the conscious mind" [36].

The methods given here were selected with the thought that you can build upon them and develop a method which will work well for you.

9

Serendipity: Invention by Accident

The word *serendipity* was coined by Horace Wolpole in 1754 to denote the gift of finding valuable or agreeable things which are not being sought. Many inventions have been the result of accidents or observations by the inventor which were quite unexpected.

According to Kivenson [37], there appears to be two ways in which invention by accident occurs. One is the situation whereby the inventor is actively engaged in problem solving but cannot find the key to get past a certain point in his progress. Fortunately, an accident or chance observation provides the solution or answer he is hunting for.

A second way in which invention by accident occurs is when the inventor discovers a new phenomenon or insight applicable to an area not related to the area of work in which he or she is actively engaged. By shifting attention to the area of the newfound information a highly successful result may be attained.

Ways to increase your chances of invention by accident will be made clear by several examples.

A CHANCE SOLUTION

The discovery of vulcanization of rubber by Charles Goodyear [38] is a classic example of the first kind of serendipity. In January 1839 he observed that rubber with sulfur added, and subjected to the proper curing by heat, is much less affected by ambient temperature variations and atmospheric degradation than is crude rubber. Goodyear was not a chemist, so he experimented with little ability to anticipate results. He relied simply on trial and error; and this procedure is closely related to accident.

Rubber had been introduced in England in the 1820s as a water sealant. Even as crude rubber it had the valuable properties of moldability, imperviousness to air and water, and the ability to undergo plastic deformation. It also had the disadvantage of becoming brittle in cold weather and soft and tacky in hot weather.

Goodyear became aware of the need to improve rubber in 1834. He had invented a valve for rubber life preservers only to learn that manufacture of that device was discontinued because of its unreliability. The unreliability was caused by the changeable characteristics of rubber.

Goodyear spent the next five years experimenting with rubber and attempting to manufacture rubber products. He had received a contract to make 150 mail bags for the U.S. government by a technique he had developed using "acid gas." After the bags were manufactured they were hung in the factory but before delivery they had turned into a sagging sticky mess. His acid gas process was successful in curing only the outer surface of the rubber.

In using the defective bags for experiments with heat Goodyear accidently dropped a piece on a hot stove. Instead of the rubber melting, as crude rubber would, it charred. He held another piece in front of the fire and part of it also charred. However, because of varying proximity to the fire, this piece had a full range of rubber from the charred to the sticky. Between the extremes was an area in which the rubber felt exactly as he wanted. At that moment he was certain that he had discovered the process now called vulcanization. The piece of material that fell on the stove was a mixture of crude rubber, sulfur, and white lead. Later it was found that the sulfur was the necessary ingredient and the white lead a catalyst.

One disadvantage in using an example from so long ago is that you may believe the days of accidental discovery are over. That is not the case. There were elements of serendipity in the discovery of penicillin, nylon, and Teflon, to name products having a modern connotation.

NEW DISCOVERIES

An example of the second kind of happy accident would be the discovery of a material which promotes healing of the inner layers of skin and mucous membrane. This happened in Cincinnati, Ohio at the Institutum Divi Thomae in the mid 1930s [39]. This research institute has been engaged in cancer research for about 50 years.

At that time the research had progressed to the point of determining that injured cells discharge hormones which stimulate the growth, breathing, and reproduction of other cells. These were called *biodynes* from the Greek *bios* (life) and *dyne* (force). A series of experiments was in progress using growth biodynes obtained from injured animal livers and respiratory biodynes from special forms of yeast. This was in a greasy base to treat skin cancers.

During related experiments one day, a flask of ether exploded seriously injuring one of the nuns. Since this greasy base material was within reach, an almost automatic reaction was to apply it to the burn. To the surprise of everyone present, the pain diminished immediately and the victim made a unexpectedly fast recovery with no remaining scar. This ointment was later used extensively by the Canadian Air Force during World War II for treatment of burns from gasoline fires after crashes. This same basic research also led to the development of Preparation H for hemorrhoid [40].

The discovery [41] of Lexan at a General Electric Research Laboratory is another example of serendipity. Lexan polycarbonate resins are a family of engineering polymers with high-impact strength and resistance to high temperature, water, and acids.

The purpose of the research program, which began in 1948, was to discover an improved insulation for the wire used to make electric motor windings. This wire, called magnet wire, must have a very thin film of insulation to conserve space. However, the insulation must resist damage as the wire is bent, twisted, stretched, or compressed during motor manufacture and must not be degraded by mechanical, chemical, electrical, and thermal stresses during the life of the motor.

By 1953 a family of polymers had been discovered which looked promising. It had proper flexibility, toughness, and resistance to high temperature but all compounds of the family were somewhat degraded by water and none had been found which could be applied to wire fast enough to be useful. During a discussion on how the resistance to water might be improved someone suggested that [41, p. 320] "it would be nice if we could start with a hydrolytically stable polymer and convert it into wire enamel rather than trying to build hydrolytic stability into a finished product."

One of the investigators on the project was Dr. D. W. Fox, who was newly employed by General Electric and was temporarily assigned to the Research Laboratory in order to become acquainted with its activities. The suggestion that a better approach might be to start with a compound known to be hydrolytically stable brought to mind an experience he had during postdoctoral work. He needed guaiacol for a particular experiment but it was not readily available. However, guaiacol carbonate was available and Fox assumed that he could put a caustic or acid with it and boil it for a few minutes to tear it apart. After several days of boiling it had not decomposed so he gave up. This seemed like an excellent material with which to begin experiments using the suggested new approach to the magnet wire insulation problem.

His first thought was to try a bisguaiacol compound but there was none available in the Research Laboratory stockroom so he chose an analogous compound among those available. It was bisphenol A, which is used in making epoxy resins. Fox describes the experiment as follows [41, p. 321]:

> I started making polymer by ester-exchange with diphenyl carbonate and the melt became more and more viscous. Eventually, I could no longer stir it. The temperature had reached about 300 °C., and I stopped at this point when the motor on the stirrer stalled. When the mass cooled down, I broke the glass off and ended up with a "mallet" made up of a semi-circular replica of the bottom of the flask with the stainless steel stirring rod sticking out of it. We kept it around the laboratory for several months as a curiosity and occasionally used it to drive nails. It was tough!

Early in 1954 the manager of the Chemical Development Department of General Electric Company's Chemical and Metallurgical Division visited the Research Laboratory and was shown the mallet. His knowledge of the glass industry and some of its needs enabled him to immediately recognize the value of this polymer which not only had phenomenal impact strength but which was also transparent. He called it "unbreakable glass." This discovery, which required two chance occurrences of the desired materials not being available, is all the more fortuitous because of the following facts:

1. Since Fox's discovery, literally thousands of other bisphenolic compounds have been tried. Yet none has been found with a better overall balance of characteristics than that produced by Fox with the original bisphenol A.
2. The accepted theories of the relationship between physical properties and chemical structure cannot explain why this polycarbonate acts as it does.

Lexan was not usable as an insulation for magnet wire. A patent issued in 1960 on an improved polyester wire enamal naming Fox as coinventor attests to success in that project. Production of Lexan as a new product to be used where high strength and transparency are needed began in 1960 at an annual rate of 5 million pounds. Among its many uses are windows and bullet-resistant barriers.

Dr. Fox is now Manager, Central Research of the Plastics Business Division of the General Electric Company.

Another example of the second type of serendipity has to do with a toy I worked on many years ago. The typical bed for a small child has sides which can be raised to a height approximately 2 ft above the mattress to prevent the child from falling even when standing in the bed or can be lowered to permit attending the child as he lies. The sides slide up and down on steel rods about 3/8 in. in diameter. Compression springs of inside diameter slightly larger than that of the rods are on these rods to prevent the sides from crashing into the lower stops with undue noise. This is shown in the partial view of Figure 9.1.

One day I noticed my 2-year-old son raise the spring, spin it around the rod, and then watch it as it worked its way down. Its diameter was large enough for it to fall freely down the rod. However, it took perhaps 5 or 10 seconds with a rotating, jerky motion I had not seen before. My son had obviously observed this earlier by the accident of a young child touching everything movable.

This movement involves a rather complex mechanical phenomenon. The inertia of the rotating member, the coefficient of friction, and the ratio of diameter of the spring rotor to that of the rod must all be within narrow ranges of values for the particular movement to occur. Observation by a stroboscope shows that the rotor proceeds down the rod by tipping back and forth on the rod as it rotates. The energy lost by friction is supplied by the change in potential energy as the rotor moves down the rod and the two must be equal to avoid having the rotor increase or decrease the speed of movements. If the unusual movement is stopped for even an instant or if the rotor slips on the rod, then the rotor falls freely. The play action of the toy I developed from this observation (see Figure 9.2) consists of starting the rotor, letting it spin down near the bottom, and then at the correct moment turning the rod end for end. This brings the rotor to the top of the newly positioned rod and so it can proceed downward again. Typically, the downward motion takes 10 seconds. If done skillfully this action can continue indefinitely. However, if the rod is turned too late, the rotor hits the end stop too hard and stops. If

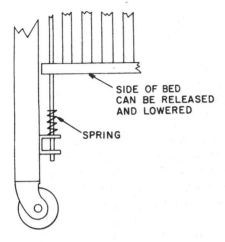

Figure 9.1 A child's crib.

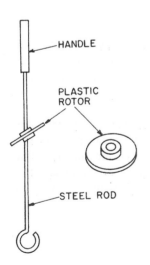

Figure 9.2 A spinning toy.

turned too early, the rotor continues upward against gravity and also stops. It should just touch the end stop as the turning of the rod is completed if the motion is to continue.

As a first step I had a search made of U.S. patents. The results indicated no previous patents. A patent application was then made. However, the patent examiner found a 1930 British patent that anticipated my invention. The first claim of the British patent (317, 404) is:

> Toy, in which a perforated disc or a ring is rotated around a circular rod-shaped body by turning impulses, characterized in that the dimensions and the weight of the disc and the diameter of the circular hole of this disc are related to the strength and diameter of the circular rod shaped body so that the inner wall of the disc, held first on the upper portion of the rod, after a single turning impulse has been given to the disc rolls on the wall of the rod with such friction that thereby the disc slides in continuous uniform rotation slowly downwards on the rod.

This and the drawing of the British patent left little doubt that it would be impossible to get a valuable patent. Inventions can end in this way or can end with a successful device. You should have no regrets if it ends as this did.

SUMMARY

It seems appropriate to make two points based upon these examples. The first is that to profit from serendipity you must take note of everything that you experience which seems strange to you, even such things as a 2-year-old child spinning a spring when you are not associated with the toy industry.

The second is that serendipity results from observation during hands-on experimentation. Invention by accident is very much more likely to occur in a laboratory or experimental hands-on environment than while developing an invention in one's mind or on paper. This indicates that the inventor who pursues experimentation actively is more apt to encounter and benefit from accidental inventions.

10

Tests to Measure Creativity

A CRITIQUE

Tests designed to measure creativity are available [42]. Typically, they attempt to determine whether or not the individual taking the test has traits which the authors of the test identify with creativity. By administering the tests correctly under correct conditions a measure of one's "creativity" may be obtained. However, the creativity measured by the test may not reflect the range of characterisitcs needed in any particular job situation. For example, at one extreme a test was developed to measure psychiatric disturbances but was then used to determine creativity on the theory that high scores on the clinical dimensions of persons who are known to be getting along in society reasonably well are less suggestive of psychopathology than of good intellect, richness and complexity of personality, and general lack of defensiveness.

Another test [43] elicits the subject's attitudes or preferences on a long list of items. Questions range from "Are you considered unconventional?" to "In doing routine chores do you often find yourself thinking of unresolved problems?" Reportedly this test was used by 283 engineers at General Motors and found to be 78% correct in identifying the creative and noncreative participants. However, the participants included engineers, engineering supervisors, research and development groups, and other personnel to whom creativity is an asset in their work situation. This brings up the question of how this heterogeneous group of participants were judged to be creative or noncreative independent of the test to determine the correctness of the results.

Other difficulties in accepting these tests are common to tests in general. Tests often measure the background of the individual more accurately than the variable they purport to measure. Intelligence quotient tests have come under criticism for this reason. For example, you will probably score higher on such tests because of the information given here regarding the tests. Other parts of the book are meant to increase your creativity as measured by your ability to invent new devices or processes. However, this discussion on tests will improve your test scores simply by making you familiar with what is expected of you.

Another aspect of all testing which must be considered is the past history of test-taking by each individual. It is usual for new graduates to take professional registration examinations soon after (or even before) graduation while they are still sharp at test-taking. High schools have been known to produce more winners of national scholarship awards by preparing students for the test. And college students make a universal practice of referring to past tests during preparation for important midterm examinations. Thus, persons who in one way or another have become familiar with creativity tests should be expected to draw higher scores than those who are not familiar.

Last, any test that is validated by considering percentages of correct results can be badly off the mark for an individual. While a large company or a government agency may be satisfied with working with gross statistics to get a general measure of the organization's ability, the individual is interested only in where he stands on the scale. Such tests can be misleading and detrimental if the results convince a participant that he is ill-fitted to be an inventor when in fact he is able to compensate for deficiencies he has in the creativity characteristics the tests measure.

TYPICAL TESTS

The information sought by creativity tests relate to many of the things discussed in previous chapters. The 310-question test published by Princeton Creative Research, Inc. and referenced earlier [42] contains statements such as:

1. It is more important for me to do what I believe is right than to try to win the approval of others.
2. I am able to stick with difficult problems over extended periods of time.
3. I often get my best ideas when doing nothing in particular.
4. I resent things being uncertain and unpredictable.

Note that the first determines your willingness to be unconventional; the second, your persistence; the third, the activity of your preconscious mind to address a problem; and the fourth, your need for routine vs. challenge. This is answered by noting your agreement with the statements, disagreement, or indecision. A scoring method produces a number which indicates six categories from exceptionally creative to noncreative.

Questions taken from other tests are

1. Do ideas excite you?
2. Do you read widely, outside your own specialty?
3. Are you careful about your manner of dress?
4. Are you annoyed by writers who go out of their way to use strange and unusual words?
5. Do you occasionally let other people overrule your opinions on other matters?
6. When faced with a problem do you try to isolate the key element on the supposition that if you do all else will fall into place?

Note that these questions all require self-evaluation. Two persons who exercise approximately equal care in dress may view themselves differently because one had parents who were fastidious and the other had parents who were careless. Likewise, being widely read probably means something different to every person taking the tests.

There are test exercises which involve performance rather than self-evaluation. One [44] is to give a set of words which are normally associated with one other word. This ability to scan the uses of the four or five words given and find the common word that intersects all of these areas of use is akin to bisociation. For example, the words elephant, bleed, lie, and wash have "white" in common association; namely white elephant, bleed white, white lie, and white wash. Note that the change of order of "bleed white" relative to the other three makes this association a bit more difficult than finding the correct word to go with bug, rest, fellow, and cover.

Another type of question requires the ability to visualize. For example, suppose you have a 3-in. cube made by stacking twenty-seven 1-in. cubes together as shown in Figure 10.1. Now suppose the outside surface of the 3-in. cube is painted red. How many of the 1-in. cubes will have no side red, two sides red, and so on?

Problems which test your ability to reorient components of the given problem or to ignore constraint which may be implied but not explicitly stated—and therefore not legitimate—test the same ability as discussed in Chapter 7. One such problem [44] is given by Figure 10.2 with the ob-

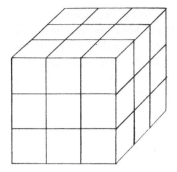

Figure 10.1 Twenty-seven 1-in. cubes.

jective of making the chain into a continuous linkage by opening and re-welding only three links. The answer becomes immediately obvious if you focus attention on the presence of three links in each chain segment rather than the four corners emphasized in the way the figure is drawn. In fact, by moving the four pieces to the orientation shown in Figure 10.3, the answer becomes obvious.

A problem [44] having to do with a self-imposed restraint is shown in Figure 10.4. Here the directions say to connect all nine dots by using only four connected straight lines. The normal thought processes leads one to assume that the lines are to be contained by the area bounded by the dots.

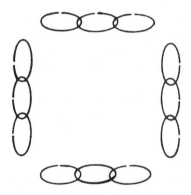

Figure 10.2 Four pieces of chain.

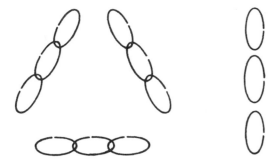

Figure 10.3   Four pieces of chain rearranged.

However, that is not stated. Figure 10.5 shows how dispelling that self-imposed constraint makes the solution possible.

WHAT IS IMPORTANT TO THE ENGINEER?

Questions of these kinds, which test performance rather than rely upon self-evaluation, seem to me to more accurately measure the characteristic of creative people. However, it is much more difficult to compose problems such as these—it takes more creativity. Thus, there is only a limited number which seem to be repeated in publication after publication. If a person has previously seen the problem or one similar to it, it measures recall rather than creativity. Furthermore, even if an engineer is good at unraveling tricks such as these, more important characteristics, in my opinion, are the intensity of his motivation to invent, his knowledge of the discipline in which he works, the breadth of his education and experi-

Figure 10.4   Nine dots to be connected by four straight lines.

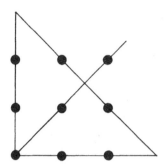

Figure 10.5 Solution by relief from nonexistent constraint.

ence, his ability to become involved in hands-on experimentation, and all the other factors discussed in Chapter 6 on improving inventing ability. A test method which can bring out all these characteristics is to suggest several actual needs related to the area of the examinee's expertise, let the examinee choose one, discuss with him or her ways to fulfill that need and evaluate the results. This is most effective if several groups of needs of increasing difficulty are presented in sequence and the procedure repeated.

11
After the Invention

THE NEED FOR RECORDS

Records should be kept throughout the time you are working on a project which could possibly result in an invention. As soon as you believe you have been successful, however, take time to review your records and evaluate them. If it seems that you have been a bit remiss, it is important that you add whatever is necessary to bring your records to completeness and clarity. However, do not add by making entries among those previously made. Rather, explain in writing that, having reviewed the development, you see the need to add information at this time. Of course, this does not give you an earlier date of invention but it does establish the information solidly as of the date of entry.

The need for record keeping and the attributes required for credible invention records are given in another book of this series [9]. In brief, records are necessary

1. To antedate a publication or an earlier filed patent which describes at least in part the invention you seek to protect
2. To antedate the invention date of another applicant for a patent on the same invention
3. To establish prior inventorship and avoid liability for infringing
4. To establish prior knowledge by you of an invention which another party claims to have given you in confidence
5. To establish knowledge of an invention prior to the employment date of a new employee who may bring a comparable idea for an invention with him
6. To determine the correct inventorship

HOW TO RECORD YOUR INVENTION

The best way to record your developments is in a notebook with bound pages. Notebooks of this type are usually available at engineering supply stores. The pages should be numbered consecutively. It should be well kept, dated, signed, and witnessed as described below.

All entries should be made in ink. If a mistake is made, line out the entry and state why it was incorrect. Fill in the page top to bottom and do not leave any blank spaces. If for some reason an unusual space occurs, cross it out with a large X so that no subsequent entries can be made. A notebook with blank spaces will be discredited as allowing entries to be made later and incorrectly dated. Enter all information in chronological order and, if possible, all entries should be made directly into the notebook.

Items such as large drawings which cannot be entered directly onto a notebook page should be signed, dated, witnessed, in the correct chronological place and put in the notebook. Photos or entries which cannot be signed directly on the slick surface should be pasted onto a page and the page signed, dated, and witnessed. You should draw lead lines across the border of the picture and into the page naming important parts of the picture to show that no subsequent picture could have been substituted.

You should enter and record all of your ideas, sketches, ways to implement your ideas, procedures, and, especially, test results. Record only factual information and keep conclusions and opinions to a minimum. Word the entries clearly so that anyone can duplicate your work without further explanation. Information about a model or prototype should be recorded, such as materials, test results, pictures, and dates on which the pictures were taken.

The notebook should always be witnessed by someone other than the inventor. Choose witnesses who are competent and impartial; they should not be coinventors or relatives. Knowledgeable friends, business associates, or professionals will make excellent witnesses providing they have the right technical background.

The witness should indicate that he witnessed and understood the material by writing "witnessed and understood by" and signing. This should be written underneath your signature and date. The witness is not a witness of your signature but a witness of the technical material involved. Make certain the witness understands and initials all of the tests which you believe constituted reduction to practice. Choose witnesses who are likely to be available later to testify that he or she did truly witness the entry.

One witness is usually enough but two are better in case one is not available at a later date. Two witnesses should definitely be used for important milestones in any development.

CONCEPTION, DILIGENCE, AND REDUCTION TO PRACTICE

In order to establish priority when another inventor has a similar patent application in process, you will need to produce evidence of the date of conception of the invention, diligence in pursuing the development of it, and the date of reduction to practice. The date of conception is the earliest date when the invention was envisioned clearly enough to be described verbally or in writing. Diligence is proven by activity, with reasonable continuation based upon your situation in making models and performing experiments necessary to perfect the invention. Reduction to practice occurs when the invention has been built and successfully demonstrated to perform as you claim it does. Keep these three important requirements of your record in mind as you ask acquaintances to witness the entries. By stating the dates by name and pointing out reasons for any gaps in your activities as you record the development of your invention, you will be able to build as strong a case as possible for the challenge which can occur.

The reduction to practice is the most likely of the three to be attacked. It is not enough that a model be built and tested; it must be given a test which reflects normal operating conditions and environment and, of course, it must perform successfully. For example, an independent inventor developed an airplane windshield de-icer and tested it in his refrigerator. It worked. Some time later he applied for a patent. In the meantime a manufacturer developed a de-icer on the same principle. He tested it under icing conditions on an airplane. His tests were performed after those of the independent inventor but before the latter's patent application was made. Both applications eventually were in process at the same time and an interference was declared. The result was that the manufacturer was awarded the patent on the basis that a static test in a refrigerator is not reduction to practice.

PRESENTING YOUR INVENTION

Something about humanity's makeup gives pleasure to discovery and invention. The normal reaction is to rush to tell the world about what you have just found. Unfortunately, that is not a very wise move. This is not to discourage writing a description and having it witnessed. Certainly do that. This negative advice has to do with your supervisor if the invention will be assigned to your employer or a prospective purchaser if you will own the patent rights. It is not meant to foster secrecy. However, a premature disclosure to those in authority to promote or reject your invention may result in your inability to describe it fully or to answer questions that may be asked about it. You may quickly find yourself in a defensive position, forcing you to guess at answers or say things that are obviously in error. It is not unusual for sufficient irritation to show through your answers under such stress that the person asking the question develops his own set of negative feelings about you and your invention.

This situation is similar to presenting a picture that is out of focus. The research of Hyman and Anderson showed [21] that once a decision is made—even if it is incorrect—more must be done to reverse the decision than would have been necessary if the decision had been held in abeyance. Thus, even if your invention is novel, useful, and nonobvious, an explanation that is less than clear, complete, and presented with confidence may result in it being rejected by the person who has the final decision.

My advice is to have a description witnessed soon after the invention occurs; but then take time to consider it from the point of view of the person or persons who will evaluate it. First of all, consider the long-term prospects of your invention. It must return the present value of all money that will be spent in putting it on the market and give promise of an attractive profit. How durable is the market for your product? Will it have national, international, or merely regional utility? How easy would it be for other manufacturers to provide the same advantages as your invention offers without infringing on your patent? How invulnerable is the new product or process to economic fluctuations?

As to projected growth in the demand for your invention, is it truly unique or an alternative design in a product area of many designs? Is demand likely to grow because of demographic changes which can be accurately predicted? Will predictable economic trends or expected changes in life style influence demand? Will there be market resistance because the invention requires a substantial investment for ancillary equipment? Will developing countries open suitable new markets?

As to marketing requirements, is the invention best distributed by a mail order business, retail stores, wholesale distributors selling to craftspersons, or sales to other manufacturers? If sold through retail stores, will there be resistance because the invention requires skills not usually available by those personnel? Will the price be set by cost plus profit or by meeting competitive prices of products already on the market? Will the product require intensive advertising? Will the invention spawn a whole line of complementary products or a single product that must stand alone?

As to engineering, will the product require any unusual laboratory facilities? What backgrounds are appropriate for the engineers and scientist needed to bring the invention to a commercial reality? Can you estimate the research and development expenditures required?

As to production, what machinery will be necessary to produce the invention? Will it require a large number of metal-forming dies, plastic molds, and welding or assembly jigs? How does the invention fit on the scale between being labor-intensive or capital-equipment-intensive?

Considering the overall concerns of a company, is the product one that is likely to improve the organization's image or have a negative effect? Are maintenance problems a concern? Will disposal after its useful life be a problem? Is product liability risk likely to be a concern?

There is no suggestion that this is an exhaustive set of questions. It is merely meant to set the tone of the thought process you should go through to learn enough about your product to present it.

The most important part of your preparation, however, is yet to be described. As answers to the above questions involve how your invention will fit into the economy, a final model reflecting the way you visualize the product or process to be in actual production should be made. You should try to learn from it just how easy or difficult your invention will be to produce. Then you should test it to determine how this prototype will respond to the requirements that will be imposed either by organizations with authority to oversee products, by the manufacturer who wishes to maintain his reputation, by the customer who wants a useful, trouble-free product, and by the environment in which the product must perform. Well armed with all of this, you will know more than anyone in the world about your invention and be able to present it with the confidence of the expert you truly are.

SELECTING A PATENT ATTORNEY

Sometime in the development of your invention the decision must be made as to whether or not to seek patent protection. This is largely an economic decision as the anticipated cost is weighed against the expected benefits. Of course, all benefits may not be monetary. Patents give status to the inventor and to the manufacturer. This must be included in the evaluation.

Assuming that the decision is made to seek patent protection, additional decisions and activities necessarily follow. First of all, you should write a full description of your invention for submission to a patent attorney. This should be a physical and functional description of each part as well as how the parts work together to accomplish what your invention does. If you are familiar with the prior art you should explain how your invention differs physically from what has been done before. Also, it will be most helpful to the patent attorney if you can describe the bisociation. What led to it and what element connected the two planes of thought? This latter information should be shared only with your patent attorney to help him write a strong patent. It must be guarded because, as stated in Chapter 5, any explanation which makes your invention sound like a logical conclusion of a sequence of steps reduces the results to engineering practice and gives grounds for an adversary to challenge the patent in court as being invalid.

The second step is to select a patent attorney. Of course, if your invention is assigned to an employer this decision will probably have been made by having in-house patent lawyers or by having dealt with a certain patent lawyer on previous occasions. Even so, you should keep the advice which follows in mind. If your invention is radically different from other products your company makes (for example, a solid state device to replace a mechanical one), a different attorney may be appropriate.

In selecting a patent attorney you must recognize that two diverse skills are necessary. First, of course, he or she must be an expert in writing patent applications and must be licensed by the U.S. Patent Office to do this work. Presumably any lawyer who is listed as a patent lawyer will have at least the minimum legal qualifications and writing skills.

The second required skill is just as important. The attorney should have expertise in the area of technology of your invention. To successfully prosecute the patent application the lawyer needs to convince the patent examiner that your invention differs from prior art by writing a clear description of it in the disclosure part of the patent application. The description I urged you to write will help but it must be rewritten into correct legal form. The patent lawyer will also need to construct the claims of your patent application to give you both the broadest and most defensible coverage possible. To do this the lawyer must have a full understanding of the scientific and engineering principles involved.

Many patent lawyers hold undergraduate degrees in one or more branches of engineering. This can provide an acceptable level of technical skill for most inventions of the corresponding technology. However, self-study or formal continued education is required as technology progresses. In large legal firms associates or partners are usually selected to provide a complete spectrum of technical expertise. Presumably, in small firms or one-person offices the patent lawyer will direct a client to someone else if he feels imcompetent to handle the case. Nonetheless, it is most appropriate that you ascertain the general nature of technological expertise of an attorney before you make an appointment by questioning the secretary and more specifically during the first part of your first meeting.

COST OF A PATENT

Three basic factors determine the cost of preparing an application for a patent: (1) government fees, which include a filing fee of $65 and an issue fee of $100 plus additional amounts depending upon the number and types of claims; (2) the cost of drawings, which normally should be prepared by a person skilled in making patent drawings in accordance with U.S. Patent Office rules; and (3) the time spent by the patent lawyer in preparing the patent application and in continued correspondence with the patent examiner during the prosecution of it.

Patent lawyer fees reflect the fact that professional people deserve compensation appropriate to their degree of skill and that the income generated must pay for all business expense. I have already discussed the need for their being skilled in both law and engineering. As to income generation, it is impossible to assign charges for every moment of a working day. Accounting for six hours on client projects out of a work day of eight is seldom exceeded. This results in only 1400 (income-producing) hours in a typical work year. Furthermore, secretarial help, office equipment, employee benefits, and so forth require about half of the income generated. You should not be surprised to learn that under these circumstances a patent attorney may charge at the rate of $50 to $75 an hour. Fortunately, with a very clear description written by you and with a model to study (both of which should be left with the attorney at your first meeting) patent attorneys are able to produce an application in reasonably short time.

Patents vary widely in length and in the prior art that must be evaluated. Roberts' patent, Quick Release for Socket Wrenches (3,208,318), described in Chapter 2, has one page of drawings and one page for the disclosure and claim. Blacks' patent on negative feedback entitled Wave Translation System (2,102,671), described in Chapter 3, has 35 pages of drawings (75 figures), 43 pages of descriptive material, and 126 claims. These represent extremes. The average patent is likely to comprise 3 pages of drawings, 6 pages of descriptive material, and 10 claims. In general as of 1980, the patentee should expect to spend on the order of $1000 on a patent which fits the "average" description and which does not have undue difficulty in prosecution. Of course, this is a proper subject to discuss with your patent lawyer during the first meeting.

CONTINUE YOUR DEVELOPMENT

The activities described as preparation for presenting your invention to prospective buyers should uncover for you the imperfections or the incompleteness of your invention as well as its saleable features. Negative characteristics would be such factors as manufacturing difficulties, excessive cost, or simply that the important invention you wish to make lies beyond what you have done. Jacob Rabinow in private correspondence points out that "one should not be carried away by the brilliance of the first version but must develop a whole portfolio of patents of other possible ways of accomplishing the same or similar results. Very often the second or third patent is more important than the first even though it is less basic." It is his experience that someone else is certain to find these opportunities for further development if the original inventor does not. His own development of the magnetic clutch discussed in Chapter 8 is a case in point. He made this invention after recognizing the low torque-handling ability of the Winslow electrostatic clutch. Furthermore, a quick check of the annual indices of the *Patent Gazette* shows that he continued to develop various forms of clutches to present a strong patent position in that device. I found, 2,575,360, Magnetic Fluid Clutch, issued November 20, 1951; 2,622,713, High-Speed Magnetic Fluid Clutch, issued December 23, 1952; and 2,629,471, Radial Flux Magnetic Fluid Clutch, issued February 24, 1953. Likewise, besides patent 2,542,430, which was shown in Chapter 7, I found 2,858,029, Self-Actuated Automatic Regulation of Timepieces, issued October 28, 1958.

The need to develop your invention beyond the first blush of success has also been emphasized in many of the examples discussed in this book. Holmes, Watt, Carlson, Gutenberg, and Kock were all said to have had previous inventions in the technology of the example, or to have developed others later, or, in the case of Gutenberg, to have continued to work in the area. Perhaps this could be said of most of the other inventors mentioned so far. However, of these I know it is true.

There is a classic case of important inventions which lay just beyond where two famous inventors stopped. It began with Edison's concern about discoloration within the incandescent lights he invented. As part of that investigation he introduced an extra electrode into the evacuated envelope and noted that a weak current flowed when that electrode was made positive with respect to the filament connections [45]. Electrons were emitted from the incandescent filament (cathode) and collected by the electrically positive electrode (anode). This is known as the Edison effect. It was not fully understood at the time and remained a curiosity for a quarter of a century.

Then, in 1904, J. Ambrose Fleming of England, who had learned of the phenomenon shortly after its discovery, performed experiments which showed that the current was unidirectional [46]. He used the device to rectify alternating current, i.e., change it to unidirectional current, and it became known as the "Fleming valve."

However, in 1907 Lee DeForest made another important invention on the device by introducing another electrode (later called a grid) between the anode and cathode [46]. This electrode was physically sparse, i.e., loosely wound wire or its equivalent. By making it electrically negative with respect to the electron-emitting electrode, DeForest found that he could reduce the current to the anode. In fact, this current was a faithful reproduction of the voltage variation applied between the grid and cathode. This was the beginning of electronic amplification.

The value of this example is that it shows how easy it is even for the great inventors to miss opportunities. In retrospect, it seems that the discovery by Edison of the phenomenon which bears his name should have touched off intense experimentation by him simply because its newness gave promise of some worthwhile development. Likewise, by the time Fleming did his work the phenomenon was better understood and the fact that a third electrode would have some effect was easy to postulate. Yet he did not take that step.

Biographies of Edison point out that the work he did played a pioneering role in heralding the electronic age. Indeed it did, but it is also true that he stopped short of the few simple steps which could have ranked him as a major contributor.

12

Preparing to Sell Your Invention

If you are fortunate enough to own the rights to your invention you will want financial gain from it. One plan you might consider is to start a business manufacturing and selling it. More will be said about this later. By far the majority of inventors who own patents want to sell their right to established businesses. The thrust of this chapter is to alert you to some of the problems involved.

SELECTING A LICENSEE

The most important fact for an inventor to keep in mind as he tries to sell his invention is that it must be compatible with the licensee's marketing, production, and engineering capacity. This is difficult advice for an inventor to follow. Specifically, the fact that the invention is an electrical device does not justify the assumption that any other electrical product manufacturer will welcome it. For most inventions you must hit the target much better than that. The point can best be made by an example. In the product area of electric switches there are residential wall switches, enclosed safety switches, bolted contact switches, and others. In general, a manufacturer of enclosed safety switches would not be interested in an invention of a new residential wall switch. A company making residential convenience outlets would be a more likely prospect. It is true that a large, diversified company may publish a catalog having both enclosed switches and wall switches, but this would undoubtedly be at different plants with no guarantee that they exchange information. There are many reasons for this separation. The residential wall switches must use high-volume manufacture techniques to sell at a competitive price whereas enclosed safety switches have smaller sales volume. The two types must comply with a different Underwriters Laboratory Standards and the marketing techniques are different. This separation among products which the novice may believe are in the same class is a typical situation. The remedy, of course, is to do the research necessary to select the group of prospective licensees which produce products that your invention will complement.

There are exceptions that can be cited when companies made the decision to change their character to seize an opportunity. This occurred when Texas Instrument decided to manufacture semiconductor devices and again when the Haloid Company of Rochester, New York decided to develop copiers based upon Carlson's patents. However, semiconductors had the prestigious research of Bell Telephone Laboratories to promote confidence of their value and xerography had the considerable effort of Carlson and Battelle Memorial Institute to justify Haloid's risk.

The reason selecting the correct company to pursue as a licensee is important is that manufacturers are very specialized organizations. A typical factory will have machinery that can do only a rather limited set of operations. Perhaps the machinery can handle metal only to a size less than your invention requires. Or perhaps your invention requires high-speed automated machinery which the company you contact does not have. On the other hand, the factory may be well equipped for your invention but the sales personnel call upon and sell through wholesale distributors and your invention must be sold as a retail item in hardware stores. As far as engineering is concerned, the company you approach may not have engineers skilled in a discipline your invention requires. Making a reliable product involves much more than having engineers who know the theory of operation. The skill of American engineering depends as much on unpublished data gained through experience by those practicing in any product line as it does the information available in books and publications which are so highly valued. Some years ago a well-known manufacturer of small motors (half-horsepower and larger) attempted to manufacture electric fan motors. The company found that the small fan motor is so sensitive to manufacturing variations such as burrs, lamination stacking, and so forth that it could not produce a reliable product. It discontinued manufacture.

INFORMATION ABOUT MANUFACTURERS

There are many ways to get information about manufacturers. A good place to start is the public or university library, which has such publications as the *Thomas Register of American Industry*. This is a multivolume set of books which lists manufacturers according to their products. One volume lists companies alphabetically with addresses and, for some, the names of executives. There are similar listings by other publishers.

Also, every industry is served by trade magazines which are even more valuable sources of information about specific companies. These magazines typically issue an annual buyers guide which list manufacturers according to the various products they make or services they offer. For example, the *1979 Precision Metal Master Directory* lists by categories those companies doing die casting, extrusion, forging, investment casting, roll forming, permanent- and split-mold casting. Then taking just one, custom forgers, it breaks the list down to closed impression dies, cold extrusion, cored, heading, no-draft, open die, powder metal, ring rolling, roll, and upsetting. Furthermore, the annual issue and the issues throughout the year earn their revenue from advertisements wherein companies tell prospective buyers of their products or services. This is of course available for your purposes.

Incidentally, if you decide that your area of expertise makes inventing in a particular line of products most likely, the trade magazines which are free to persons qualified to receive them by virtue of their jobs in industry are also available to anyone else for a modest subscription fee. Receiving the publication over a period of time will enable you to learn much more about the industry, the companies within the industry, and even some of the personnel whose names may be of more value than will a short-term research project done while you are anxious to present your invention to prospective buyers.

Other sources of information about a given industry are the trade organizations. Those whose members participate in the writing of standards typically list all member companies which make the product to which the standard applies. Furthermore, a directory of member companies is usually available. Chapter 15 gives advice on how to locate standards for a given product and this same technique can lead you to the manufacturers of that product.

Finally, for all publicly owned companies a very detailed source of information is the annual report. This may be available at the offices of a local stock broker. Copies can often be obtained by writing to the company. These reports not only contain financial information but also explain company operations and often show new products introduced during the preceding year.

LETTER TO PROSPECTIVE LICENSEE

In writing to a prospective licensee try to put yourself in his place and anticipate the questions he would most likely have about your invention. The letter should be brief and should include the following:

1. That you have an invention which is an improvement on a product or which is a totally new product. Name the product.
2. What the improvement or new product does but not how it does it.
3. A brief statement of the patent protection you have, e.g., a patent has issued (give number) or a patent application is pending and so forth. Do *not* give the patent application number if the patent has not been issued!
4. A statement which explains why you believe your invention will fit well within the company's product line (based upon your research of possible buyers). If what you say is true the executive who receives your letter will know you are competent and should be taken seriously.
5. A statement that you are willing to sign a nonconfidential disclosure agreement and wish to submit information about your invention by meeting with appropriate personnel or by mailing a detailed description of the invention to them.

Assuming a favorable response, you should realize that you probably will have but one chance to sell your invention. If the prospective buyer is not convinced expect nothing more than a statement that the company does not wish to pursue the matter further. They are not interested in debates or rebuttals on the matter. If you have done research on the company and on competitive products you will be in a position to give your best effort on the first information submittal. There should be no need for a second chance.

If a client company is interested enough to discuss an agreement, you would do well to engage a skilled negotiator with legal expertise to negotiate. You will have spoken to him about the matter and be able to give his name immediately. (Patent attorneys usually offer this service.) Having an intermediary will not only help to avoid ill feeling between you and the company personnel with whom you need to work to bring the invention to production, but it will also lessen the possibility of the compensation being inadequate.

If there is no positive response after a number of tries you should pause and analyze the situation again. Is your invention deficient in some important characteristic which your contacts have recognized? Does the invention require skills those companies do not have? Would that be true of all companies? If so, you may find the only alternative you have is to manufacture it yourself or accept the fact that it probably will not sell. The first alternative takes much effort but many fortunes have been made by inventors proving the worth of their inventions by starting their own companies.

THE NONCONFIDENTIAL DISCLOSURE AGREEMENT

You may feel a bit overwhelmed by the company you approach, but the company, in turn, feels that it must protect itself against you. There is a legal right to restrict other parties from using information that you disclose in confidence. An inventor may describe a device very similar to the one a prospective buyer of his invention is developing. Then, instead of relying on patent right determined by who is first to reduce the invention to practice and whether or not due diligence was exercised, he may claim a right to compensation based upon "confidential disclosure." To avoid this, prospective buyers usually require the inventor to sign a release from any obligation except that which is derived from patent rights the inventor has or might receive at some future date. The agreement can be stated in many different ways. An example is shown on the next page.

THE XYZ CORPORATION
101 Opportunity Way
Pleasant Town, Any State

TO THE OUTSIDE INVENTOR: STATEMENT OF POSITION

The XYZ Corporation is constantly striving to improve its products and expand its business. Therefore, it is willing to consider the ideas which outsiders submit to it, but only on the terms set forth herein:

(1) Submission of an idea or invention, patented or unpatented, is gratuitous, creates no confidential relationship with the company and does not obligate the company to accept or pay for the idea or invention, or to pay any expenses the inventor may incur in making the submission.

(2) If the company considers the idea or invention new, and useful to the company, the inventor will be so informed in writing and the company will attempt to negotiate with the inventor a mutually satisfactory written agreement providing for compensation to the inventor.

(3) The company will not make any payment, or undertake any obligation to pay, for any submitted idea or invention except pursuant to such negotiated written agreement.

(4) The company in any event reserves the right to retain all materials submitted to it for consideration, in order to preserve an accurate record of the idea or invention submitted and the date of submission.

If you find the foregoing conditions acceptable, please execute the attached form of Release and return it to the company, whereupon the company will give your idea or invention such consideration as it consideres justifiable.

RELEASE

I have read and understand the above statement. In consideration of The XYZ Corporation's examining my suggestion or invention pertaining to the following:

_____

_____

as set forth in the following submitted materials: (enumerate letters, drawings, photographs, etc.) _____

_____

I hereby release The XYZ Corporation from any obligation or liability with respect to any information submitted to it relating to my idea, improvement or invention, except such obligation or liability as may arise from any written agreement that may hereafter be entered into between us, or as may arise from infringement by it of any valid patent that I may obtain covering my invention.

_____
(Signature)

_____   _____
(Date)   (Address)

13

Compensation

There are two different categories of inventors. In one group are those inventors who have been employed to do product development. In that situation, any patent having to do with the company's business must be assigned to the employer. This is true even if the invention was purposely made away from the inventor's place of business. The rationale is that the invention resulted from or at least was influenced by information acquired on the job. Many employers have engineering personnel sign a statement acknowledging the company's rights in this matter but even in the absence of such an agreement the courts have held that accepting a salary from the employer gives an implied contract.

In fact, if you make an invention which has nothing to do with your employer's business but you do it on company time, using company equipment or using material from company stock, the company can demand the patent rights. The only safe way to retain patent rights if you are an engineer employed in product design and development is to invent something that has nothing to do with the employer's business and be able to prove that you did not use your employer's resources in any way. Even so, controversy can arise as other divisions of your company engage in new product lines or new divisions are acquired. The safest procedure is to define the product areas in which you wish to work and ask your employer for a signed release for inventions that may result.

It should be made clear that "patent rights" means the monopoly granted by the patent. If you are the inventor you must be listed as such. The right to be publicly acknowledged as the inventor cannot be taken from you under any circumstance.

WHEN YOUR EMPLOYER HAS PATENT RIGHTS

It is appropriate that the compensation an employed inventor can expect if his contribution results in a patent be discussed. Konold et al [9] includes a chapter on licensing inventions and that information will not be repeated here. Rather, the main thrust of this discussion will be to give some idea of what constitutes reasonable compensation.

The compensation received by an engineer-inventor from an employer may not be well understood. Often there is no special transfer of money designated as compensation for the invention. Yet the employed inventor is well compensated. The employer must recognize the employee's value because the patent will alert every other employer of the industry that here is someone special. The inventor's address is also given on the patent. Thus, there is no difficulty in his being contacted by other employers looking for a creative problem solver. Every employer knows this or soon learns. As an example, Rabinow states that after his invention of the magnetic clutch, which had to be assigned to the U.S. government, he received a raise that made his salary larger than that of his supervisor.

Not all inventions create the excitement of the magnetic clutch, but even a modest raise when viewed in terms of total income over the remainder of your working career is likely to be fair compensation for your contribution. A truly creative engineer will certainly recognize more than one inventive challenge and, thus, increase his or her earnings appreciably. The reason most engineers do not invent is that they do not seek out the opportunities to do so; it is not that the opportunities are not there or that they cannot invent.

Besides the monetary rewards, the inventor receives other important compensations. At the moment of bisociation he will experience a satisfaction that is likely to be remembered for life. Archimedes shouting ''eureka!'' as he ran from the bath is not an exaggeration of the elation an inventor feels. He will gain status and respect from his associates and, because inventors are special, job security. True, one cannot measure these less tangible compensations, but they are there and they are valuable.

WHEN YOU HAVE PATENT RIGHTS

The other category is that where the inventor owns the rights granted by the patent. He will probably be interested in selling the rights. Selling patent rights is perhaps the last vestige of true free enterprise. Any arrangement that is satisfactory to both parties can be worked into the agreement. Perhaps the main point is that two factors will influence the rewards that the inventor can demand. One is the recognizable worth of the invention to the buyer. The other is the reputation the inventor has gained from past inventions which resulted in successful products. The tendency, unfortunately, is to think so much of your invention that demands are excessive. Realize that the prospective buyer will evaluate your patent on the basis of increase in profits it will bring to the company. You should make a realistic evaluation on the same basis. Will your invention bring about lower manufacturing cost? Will it increase the demand for the product? Will it start a whole new industry? Is the patent likely to withstand a court challenge? The answer to questions such as these is the starting point in deciding how much your patent rights are worth. The successful freelance inventors whom I know agreed to modest compensation for early inventions and used these successes on which to build a reputation.

As a second step the inventor should consider three types of transactions: the outright sale, the exclusive license, and the nonexclusive license. A transaction somewhere in the middle is a limited, nonexclusive license, i.e., inviting only a selected number of licensees into the "club" and imposing upon them some conditions that are not unlike those imposed on the exclusive licensee in order to maintain membership in the club.

Tax considerations are involved. All substantial rights, either by an outright sale or an exclusive license, must be sold in order to take advantage of reporting income as capital gains. The capital gain feature may be so important, perhaps depending upon the inventor's income bracket, that it dictates to some extent the manner in which the patent rights should be sold. The amount of compensation obviously varies with the demand for the invention on the part of both the prospective licensee and the consuming public. But in any given demand situation various types of approaches to compensation may be satisfactory to both licensor and licensee.

As negotiations start it may be advisable to give the licensee an option for a specified amount of money (say a few or many thousands of dollars) for a limited period of time simply to consider whether or not the licensee wants to take the license. The option time would be for the purpose of either a market research or a technical study. Its advantage to the prospective licensee is that the time and money spent on a careful evaluation of the patent will not be lost by sale to another interested party.

The licensee may be persuaded to pay money "up front" as a lump sum advance against royalties or simply as a lump sum down payment which is not an advance against royalties but with the understanding that royalties will be in addition to it. The amount of the lump sum may depend in part on the demand of the invention and may depend in part on the continuing royalties. For example, one inventor offers alternatives for the nonexclusive licensee along the following lines:

$4000 advance against royalties and a 6% royalty
$8000 advance and a 5% royalty, etc.

The prospect thus has the choice of putting a lot of money up front which is helpful to the inventor and opting for a lesser royalty, or, alternatively, putting less money up front but paying a much larger royalty. The up-front money has an advantage of forcing the licensee to take an active role in promoting the invention. If it costs nothing to get a license, particularly a nonexclusive license, perhaps the licensee will do little to launch the invention.

Sliding scales of royalties based on production have their advantages. An increasing royalty based on increased sales has been used. The idea there is that the licensee needs a low-royalty situation to get the product launched. The company should be spending money early on product development, advertising, and the like. However, once the product is launched and is enjoying success, then the royalty should be higher in order to compensate the patentee for giving the invention away at a low royalty in the early stages. The opposite sliding scale has also been used. A higher royalty in the beginning with a low royalty after increased sales encourages the licensee to increase sales effort in order to get the benefit of the low royalty. This method is most often used on low-volume, high-technology products with unusually high initial royalty.

A minimum annual royalty is usually used in order to protect the patentee in the case of an *exclusive* license, but is not normally used in the case of a nonexclusive license. In the case of an exclusive license, it is obviously important in order to assure that the licensee will not simply sit on the invention. It also gives leverage to force marketing activity of the invention.

Royalty rates range all over the lot. The lowest rate I know of in reference to an actual case was about 0.05%. The product, however, was an electronic organ and 0.05% (or whatever the royalty might have been) involved a substantial amount of money in return for a circuit constituting only a small part of the organ. Similarly, employees at the Crosley Corporation when that company had exclusive rights to use shelves on the inside of refrigerator doors (circa 1946) spoke of the inventor receiving $0.50 per refrigerator. That amounted to only 0.20% of the retail price of the refrigerator but the volume of a quarter million units per year made the total compensation a handsome amount indeed.

The largest royalty I have heard about was 20%. A 20% royalty would be paid only where the product was so unique and the protection so good that the licensee would not have to be concerned about competition on the product.

The competitive factor unquestionably affects the amount of royalty that the licensee can pay. If a similar product can be manufactured free from royalty, then the licensee might be put in an unfavorable competitive position by having to pay too high a royalty.

14

Refuting the NIH Theory

Selling ideas for product improvement or patents is difficult. A popular way to explain failure in these endeavors is to allude to the not-invented-here (NIH) theory. As usually explained this theory holds that engineers and company executives while seeming to welcome independent inventors do everything they can to thwart the use of inventions made by anyone other than the company's engineers. An article in *Spectrum* (47) gave evidence of how firmly this theory is established. The first part of that article states [47, p. 44]:

> Once the main source of new U.S. patents, the independent inventor's output has been dropping. There is some concern that if he is not actually an endangered species, he is an increasingly discouraged one for whom such perennial problems as the resistance of corporate project engineers to outside ideas have been compounded by the trend toward "Big Science," the increased cost of patent litigation and other factors.

Later in the same article a section is headed "Engineers as invention killers" in bold type. There the statement is made that "this guy (the corporate engineer) will do everything he can to kill the project."

There is no doubt that everyone, including every corporate engineer, works on his own ideas with more enthusiasm than on someone else's ideas; but this is true in every endeavor. Consultants in advertising, personnel relations, marketing, industrial engineering, etc., have this problem to face as much as do inventors. The frequent reference to the NIH theory implies that there is an obstructive attitude that goes far beyond the lessening of motivation that everyone experiences in working on a project with which he is not emotionally involved.

Invention is so important to the national well-being that causes of the dwindling output of the independent inventor must be explored. Yet this is not the type of problem that lends itself to the controlled experiments of scientific research. At best, opinions formulated from experiences and interest must be stated and given the test of counteropinions. It is in this vein that this rebuttal of the NIH theory is written.

My interest in invention has been intense during the past 25 years of teaching courses in product design. As an inventor, the experiences of being flown to Cleveland, Chicago, etc. with the hope of a contract at the other end have been known. And so have the disappointments of the polite regrets. More recently the bulk of experience has been at the other side of the table, in making decisions of whether or not to recommend that a contract be offered an inventor. Thus, this discussion reflects what has been learned from seeing both sides of the process.

There are other factors, more important than the NIH theory, that have reduced the output of the independent inventor during the past quarter century. These factors are exposed with the hope that understanding them can be a first step in helping the independent inventor increase his valuable contributions. They are

1. The proliferation of standards and the increase of importance attained by them during the last 25 years
2. The affluence enjoyed in this country during that same period with the concomitant lessening of motivation for that segment of the population from which independent inventors come.

THE EFFECT OF STANDARDS

During this time period American industry has witnessed a dramatic increase in the number, the extent and the authority of standards. These standards impose constraints that can be as fatal to a new idea as violating a principle of physics. Thus the dilemma. On the one hand standards promote efficiency in production and increase safety. I have discussed their value previously [48] and nothing stated here is meant to argue against their continued development. Especially useful are the voluntary standards developed by persons of experience in the field covered by each standard. However, standards do impose constraints and increase the risk assumed in making changes. It must be recognized that the American public cannot have the best of both worlds. Standards always reflect what has been done; they inhibit what might be done.

How do standards thwart invention? Principally, by imposing an area of ignorance on most would-be inventors. It is unusual for an inventor to spend much effort on a fledgling idea without some investigation as to whether it is novel, useful, and non obvious—the three tests for patent ability. However, an idea that passes those three tests well may not pass the tests imposed by industry because it may run counter to a standard which dominates the field of the invention. In fact, the lack of a previous patent may very well be caused by knowledge of those skilled in the field that such a design would be rejected by a standard. Precisely because a void exists, a less knowledgeable inventor can secure a patent on such a device. Patent examiners rightfully do not take existing standards into account when determining patentability. The successful inventor needs to compensate for this Patent Office procedure by becoming as familiar with the standards that apply to products which could result from his invention as he is with the prior art in the area of his invention. This is frustrating, detailed "paper" work that a freewheeling idea man usually eschews.

A few examples may substantiate this argument. The simple wall switch used in every home to turn lights on and off is given an endurance test of 30,000 operations by Underwriters Laboratories Inc. (UL) as a prerequisite for listing by them. Failure to pass the endurance test will effectively eliminate sale of the device. Thus, a manufacturer will reject an invention that gives even a slight increase in probability of failure even though it has other advantages.

Some years ago an inventor received several patents on what he described as a heavy-duty, high-current enclosed safety switch. Indeed it was heavy duty. The quick-make, quick-break mechanism which is usually made of sheet steel parts were made of steel castings. This design reflected his experience and considerable expertise in the foundry industry. The internal-external cam arrangement he used had quite a bit of friction but it provided a large mechanical advantage. The main problem with the design was that the mechanism was judged to provide insufficient break distance to snuff the arc which occurs during operation on the horsepower overload tests prescribed by Underwriters Laboratories and the National Electrical Manufacturers Association. The point is that this inventor would have increased his probability of success by studying several key standards while he was developing his invention. Or he would have found it impossible to provide the needed break distance, which was my judgment, and would have gone on to something else.

As a final example, consider the present situation with respect to the rotary lawn mower. Trade journals have quoted leaders of that industry as expressing doubts that the new standard sponsored by the Consumers Product Safety Commission (CPSC) will permit the product to even survive. With such pessimistic evaluations of the design constraints that industry must face, the folly of attempting to make improvements on that product without thorough study of those standards should be obvious.

It is difficult to estimate the number of standards that apply to American products. Certainly they number in the thousands. Very few products can be manufactured completely untouched by standards. Underwriters Laboratories Inc., the National Electrical Manufacturers Association (NEMA), the National Fire Protection Association, the American Society for Testing Materials, and the Institute of Electrical and Electronic Engineers are among the most important standard writing bodies for electrical products. A similar list could be written for every other engineering discipline. There are products which simply have no market unless one or more of the standards of these groups are met.

Recently, the federal government has been active in imposing standards through OSHA and CPSC. The latter is currently sponsoring standards on products for which standards already exist but which are judged by these bodies to be inadequate.

Not all standards are exposed by an official designation as such. Organizations such as Consumer's Research effectively impose informal standards by establishing criteria that products should meet to gain a favorable review in their reports to subscribers. Since these organizations do not publish standards their criteria can only be known by reading evaluations of similar products and forming a judgment of what each evaluating organization would demand of a new product.

Standards vary in authority from those that are imposed with force of law to those that have the character of suggestions. No matter what the status of a particular standard is, a manufacturer may reject an idea that runs counter to it. The rejection is a result of a defect in the invention, not a result of the NIH syndrome.

THE LOSS OF MOTIVATION

The general condition of prosperity over the last quarter century also has been counterproductive of independent invention in a number of ways. First of all, creative people have been able to find jobs of sufficient challenge to be both financially rewarding and personally satisfying. It is indeed difficult for a would-be inventor to turn his back on the "safe" opportunities available to him in favor of the high risk of failure as an inventor. Special situations can, of course, exist. For example, one very successful engineer, in response to foresight which convinced him in the middle 1960s that the long boom in the aerospace industry was about to end, took on inventing as a second occupation. He used his engineering know-how to develop a delightful toy that has been a staple in the industry for over 15 years. Had the aerospace industry continued to give promise of advancement his story might have been much different. In this case insecurity of one industry provided the motivation.

Perhaps the most counterproductive ramification of prosperity is that it is a major thief of time. Whereas the optimist will correctly say that the leisure industry had created many opportunities for invention, an equally realistic evaluation is that many who in former years would have found satisfaction in pursuing an idea of a new product during evenings and weekends find the pull of television, camping, and participative or spectator sports too attractive to resist.

The reduced motivation resulting from prosperity has also had a less obvious effect on the work of the inventor than directing leisure time elsewhere. The whole country seems impatient for rewards. Hard work and sacrifice seem too out of step to discuss. Certainly it would be nice to receive royalties for merely suggesting an idea but few inventions warrant that. Yet many inventors wish only to present an idea. Often this is in the form of a sketch or a crude model. Case studies of inventions that have been highly successful show that usually development of a device or process took enormous effort and that it was ready for manufacture before the invention was accepted. The need for a professional approach to the presentation of inventions to prospective buyers was discussed in Chapter 11 but deserves amplification here.

Two contrasting examples will help make this point. The first is the well-known invention of xerography previously discussed. Carlson had approached a long list of manufacturers he judged to be likely purchasers of his rights [12]. The quality of his early copies, however, left much to be desired. It was only after he spent many years and considerable money and finally assigned a large portion of his rights to Battelle Research Institute for help in perfecting the process that he was able to sell his patent rights.

The contrasting story starts with a letter received by an electric switch manufacturer which stated that a new switch had been invented and a machine developed to produce them at 1800 per hour. Would the company be interested? The answer, of course, was yes. After the usual safeguards about disclosure, a request was made to see the device and drawing of the machine. Finally, drawings of a sort were sent with such little detail that they could not be interpreted. Further correspondence regarding a model produced one that did not work and the admission that the machine did not produce the switch but merely formed a leaf spring used in the switch. If the promise made in the first letter had been kept there would have been no trace of the not-invented-here syndrome. However, the evidence in hand after much correspondence was that the leaf spring was so overstressed during operation as to be useless and that a machine to produce 1800 units per hour had never been built. Perhaps a major development effort would have salvaged the product but few ideas are worth a major development effort on the part of the purchaser of the patent rights.

OTHER FACTORS

The list of factors most influential in lessening the output of the independent inventor was not exhaustive. The economics of the proposed process or device, its compatibility with the manufacturing capability and the sales effort of the company, anticipated acceptance by the consumer, increase or decrease in product liability exposure, and many other factors are also important. An adverse assessment of any one of these probably will bring a negative response and none have anything to do with not being invented here. The mere novelty of an idea does not make it valuable. This is a necessary but not a sufficient reason for a company to offer a contract. The inventor must take the responsibility of presenting a product that meets the same criteria as does every other product a company produces. The model must make it obvious that the invention is completely worked out and ready for production. This will do more to sell inventions than any other single factor.

America needs inventions today as much as ever before. Inventions to increase productivity are perhaps the only way to return the country to the rapid improvement in the standard of living that was taken for granted just a few years ago. The independent inventor, fired up with an overwhelming desire to win his spurs, is our greatest hope.

15

Finding Standards that Affect Your Invention

Standards were mentioned in Chapter 1 and they were an important part of the argument in the chapter just concluded. It seems appropriate, therefore, that you be given some guidance as to what standards exist and how to locate those that are important to you.

TYPES OF STANDARDS

There are many kinds of standards. There are workplace standards, product standards, mandatory standards, and voluntary standards.

A workplace standard sets the rules and regulations concerning the total environment of an employee when he is actively working on the job, while a product standard is one designed to encompass the requirements for a specific product. The standards developed by the Occupational Safety and Health Administration (OSHA) are workplace standards, but a product manufacturer cannot obtain approval of product design through OSHA. In some cases, products can be submitted to independent testing laboratories to determine compliance with workplace standards.

A mandatory standard is one having the force of law. For example, the various standards adopted by OSHA and published in the *Federal Register* are mandatory standards. A voluntary standard, on the other hand, is usually developed and promulgated by a trade organization or similar group. Compliance with a voluntary standard is not a requirement of law, even though noncompliance can often result in serious legal problems.

The concern here is almost exclusively with product standards. Not all such standards are necessarily the same in either content or authority. Thus, although this section deals specifically with safety standards, it should be noted that voluntary types of standards also exist which deal with such things as interchangeability of parts.

The requirements imposed upon a product by a standard or a code can be written in only two ways: as prescriptive requirements or as performance requirements. A prescriptive requirement deals with materials and dimensions. For example, according to the Underwriters Laboratories Standard for Cabinets and Boxes (UL50), steel enclosures for electrical equipment to be used outdoors must be protected by a zinc coating defined as G90 by the American Society for Testing and Materials (ASTM) or by a G60 coating and certain types of enamel. As another example, Underwriters Laboratories Standard for Enclosed Switches (UL98) requires that the distance between uninsulated parts of opposite polarity be ¾ in. through air and 1½ in. over surface for voltages between 126 and 250 V. These are called prescriptive requirements because, in effect, they determine the design decisions rather than permiting the engineer to decide what is adequate.

A performance requirement, on the other hand, deals with a performance test that the product must successfully complete in order to comply with the standard. These typically involve increasing the electrical, mechanical, or thermal stress on the product and then operating it repeatedly. The combination of increased stress and number of operations is designed to represent worst case use or to accelerate wear-out. For example, representative circuit breakers that are rated at 15 A and are of the type used in homes to protect the wires running to the convenience outlets are required by Underwriters Laboratories to be able to interrupt rated current 6000 times, interrupt 6 times rated current 50 times, then interrupt a current that would reach 10,000 A if the breaker were not there 3 times. It must still be operative after these and the 9 or 10 other tests that make up the evaluation program.

Performance requirements allow the inventor full latitude in finding ways to comply, and from that point of view they promote creative design and continual product improvement. They are preferred over prescriptive requirements. However, most standards contain both prescriptive and performance requirements.

STANDARDS ORGANIZATIONS

As an inventor, you will need to locate standards that might apply to your product. I will endeavor to give you sufficient information in this section to start your search. However, it is not possible to give a complete and definitive list. You should be wary of any publication that implies that it does this. The number of available standards in any product area is large and continues to grow. Often the origins are organizations that previously have not been engaged in standards work. Their publications might come as surprises but once available they cannot be ignored.

On the other hand, the problem of acquiring necessary standards is not overwhelming. There are relatively few organizations that account for the majority of standards. Careful surveillance of these organizations, participation in professional society activities, and affiliation with trade associations appropriate for the class of products of interest to you will probably put you in touch with the standards you need.

Another important source of standards information is the *Product Standards Index,* Second Edition, by V. L. Roberts, published by Pergamon Press. This 500-page book identifies many standards-writing organizations and lists standards that apply to various products. Also, an *Index of Federal Specifications and Standards* is available from the Superintendent of Documents, U.S. Government Printing Office, Washington, D.C. 20402.

The following organizations can be considered as primary sources of standards.

1.   The American National Standards Institute (ANSI) does not write standards. Its purpose is to be the clearinghouse for standards that are written with full participation of all concerned. Many of the standards of other U.S. organizations, as well as foreign standards, have been designated as ANSI standards. Thus, this organization is the first place to look for standards that might apply to a given product.

2.   The Underwriters Laboratories (UL) is probably the most widely known standards-writing and testing organization in the United States. Their standards apply to materials and devices and are meant to prevent loss of life and property from fire, crime, and casualty. There are over 350 UL standards, three major testing laboratories (New York, Chicago, and Santa Clara), and offices of local inspectors throughout the United States to provide constant surveillance of products before they leave the factories where they are produced.

Products that comply with UL standards bear a label or the initials UL circumscribed by a circle as evidence of compliance.

3.   The American Society for Testing and Materials (ASTM) develops standards on the characteristics and performance of materials. It publishes over 4000 individual standards in over 30 volumes, and each standard is also available as a separate publication.

A typical ASTM standard identifies a certain property of a material that is of interest, describes the equipment necessary to carry out the test that quantifies the property, and carefully describes the test procedure.

4. The National Fire Protection Association (NFPA) promotes and improves methods of fire prevention and protection. The NFPA publication list is extensive and touches any subject that could conceivably be associated with fire. The *National Electrical Code (NEC),* one of its publications, is accepted as the standard for installation of electrical equipment by most political subdivisions throughout the country and, as such, is the source of many product changes that are eventually made in UL standards. *NEC* is reissued every third year, with changes agreed upon by the members of 21 panels in response to suggestions submitted to them.

Also, the NFPA's authoritative *Fire Protection Handbook,* now in its 14th edition, and their *Fire Protection Guide on Hazardous Materials* are of special interest to engineers concerned with product improvement.

5. The National Safety Council (NSC) devotes its entire effort to the prevention of accidents. Part of its effort is directed toward general educational activities, as indicated by the recent involvement with the National Electrical Manufacturers Association (NEMA) in developing slide tape presentations on the use of ground fault interrupters to prevent electrical shock. However, most of its publications are related to the design of chemical plants.

There are certain nonmilitary government agencies that write standards or solicit and support standards written by organizations that may be interested in doing so. The activities of these organizations should be watched carefully because these government agencies are given the power by Congress to impose their standards with legal action.

6. The Food and Drug Administration (FDA) is the oldest consumer safety and protection agency. Its origin was the Food and Drug Act of 1906, but today its activities also include cosmetic products. This agency endeavors to use voluntary compliance, but it can rely on court orders to seize a dangerous product. The inventor's involvement with this agency is usually related to food- and drug-processing equipment or containers.

7. The Federal Trade Commission (FTC) is related to product restrictions mainly through the Magnuson-Moss Warranty Act. It also

warns the public when it believes a product to be unsafe or dangerous. This has happened in recent years when Christmas tree lights and dolls with eyes of poisonous seeds were reported.

8.   The Occupational Safety and Health Act of 1971, which established OSHA, adopts consensus standards such as the NEC or sets its own standards and determines if workplaces comply with these standards. Its inspectors levy penalties if violations are found. Repeated violations can result in heavy fines and jail sentences. Obviously, any citation issued by OSHA which implies that your invention is unsafe would be commercially damaging. Thus, anyone involved in the development of products that are likely to be used in industrial installations should know what standards OSHA relies upon in that product area to be certain of full compliance.

9.   The Consumer Product Safety Commission (CPSC) was created in 1972 to protect the public from unreasonable risk of injury from consumer products. It is primarily concerned with products used in the home. The first activity of the agency was to determine which product categories most frequently cause injury. They found that "certain makes and models of 16 categories of products subject the consuming public to unreasonable hazard." These categories are, in alphabetical order, architectural glass used in sliding doors, color television sets, fireworks, floor furnaces, glass bottles, high-rise handlebar bicycles with elongated seats, hot-water vaporizers, household chemicals, infant furniture, ladders, power tools, protective headgear, rotary lawn mowers, toys, unvented gas heaters, and wringer-washers.

One of the problems the CPSC has is to determine just what constitutes a consumer product. The key to distinguishing a consumer product is its ability to provide its function when standing alone. A fire alarm functions independently of being mounted in any particular place and is considered a consumer product even when built into a home by the builder. On the other hand, roofing shingles serve no purpose until properly attached to the roof by the builder and are thus not a consumer product.

10. The Environmental Protection Agency (EPA) was created in 1970 to coordinate federal environmental activities. Authority is given to the agency by the Clean Air Act; Water Pollution Control Act; Safe Drinking Water Act; Solid Waste Disposal Act; Federal Insecticide, Fungicide and Rodentcide Act; Toxic Substances Control Act; and Noise Control Act. The EPA sets environmental quality standards, monitors pollution levels, and sponsors research related to environ-

mental pollution. Its requirements on automobile exhaust pollution are well known to the general public and have been a major influence in automobile design. Its involvement with most companies, however, is to reduce the objectionable effluents from the manufacturing facility. Thus, chemical engineers are most likely to be the ones involved with this agency.

The following two important consumer research groups should be known to engineers involved with consumer products:

11.   Consumers Research, Washington, N.J. 07882 and Consumer Reports, Orangeburg, N.Y. 10962 publish periodic reports that discuss the quality of products they purchase from retail outlets, i.e., not specially prepared samples. The reports typically include a general discussion of what they believe to be important for customer safety and satisfaction as well as ratings on how the specific devices they tested performed.

SUMMARY

Standards have long been important to the engineer and are even more important to the inventor. You should write to those organizations which are likely to have standards in your product area. Ask for a catalog of their publications. They usually provide this at no charge because selling standards is an important source of their income. The study of these catalogs should give you the titles for an adequate start for your library of standards.

Ask also for any publication offered on the history and operation of the organization. Learn how each organization arrives at its standards, what enforcement authority it has, and how compliance with its standard is verified. This information is most important if you decide that your improvement is incompatible with the existing standards.

16
Case Histories

Throughout this study of inventing a primary source of information has been real life examples. Much can be learned in this way about the kind of person who invents and how inventions occur. This is not to suggest that you imitate the examples; everyone must find the way that best fits his talents and circumstances. Nonetheless, in this chapter three short case studies have been selected to fit the broad categories of readers, namely, scientists, mature engineers, and engineering students. Perhaps you can see yourself acting in much the same way as these three inventors.

A LIFETIME OF INVENTING

John F. Dreyer was graduated from the Massachusetts Institute of Technology in 1929 with the degree of Bachelor of Science in Mechanical Engineering. He was employed by Formica from 1929 to 1945 as a development engineer on high-pressure laminates. While there he developed a luminous laminate which was widely used as instrument panel material in military aircraft. In 1945 he decided to start a company doing research and product development related to the transmission of light. He spent the next year doing research in the basement of his home.

I met John around 1950 when, as a young instructor at the University of Cincinnati, I was asked to join a group doing research on liquid crystals* in the Chemistry Department. At that time he had already founded

---

*Materials are called liquid crystals if between liquid and solid states there exists a phase wherein the molecules conglomerate and take on the order of crystal structure while still having the fluid characteristic of a liquid.

the Polacoat Corporation and acted as President and Director of Research. He had developed a technique of rubbing glass or clear plastic, spraying it with dyes composed of long-chain molecules, and thereby producing light filters that eliminate one component of vibration of visible light. These are called polarizers. One use was to protect personnel from eye damage during test atomic explosions. Later this same invention was the basis for manufacture of the special eyeglasses needed to view three-dimensional movies.

John has a long history of awards. In 1940, he received the National Annual Plastics Product Award for development of a laminated Pregwood aircraft propeller. In 1976 he received an honorary Doctor of Science degree from Kent State University for his assistance in the development of their Liquid Crystal Institute. In 1978 he was named Distinguished Scientist by the Technical and Scientific Societies Council of Cincinnati. John holds 28 U.S. patents and 19 foreign patents, most of which list him as sole inventor.

I asked John to tell us about his invention of the polarizing light filter. This is his story.

On the wall of the reception room at the U.S. Patent Office in Washington, D.C. there used to be a sign:

"Don't just sit there, invent something"

The development of a light-polarizing filter is an example of the steps through which an invention can progress.

The initial motivation was the desire to get into business for myself by using what I though were my best talents, perseverance and a strong imagination. I was convinced that in this way my success would truly be dependent only on my own ability.

The time was 1931, when an idea came. Television was just beginning by using spinning lenses. I though why not use the marvelous disciplined arrangement of molecules in a crystal as a reference base for creating an image. This is not a simple idea but it was a place to start. The perfect lattice structure of a crystal, if put under control electrically, could become a marvelous device. The idea included using the piezoelectric property of crystals to give the link between the crystal pattern and an applied electric field. However, I did not know how I would make the pattern visible.

This optical problem was chosen for the first study. All available evening and weekend time was spent in the local libraries even though I was employed in a research department in an entirely different field.

The initial investigation was on all means of producing and controlling light. Each method, fluorescence, phosphorescence, incandescence, electroluminescence, chemiluminescence, and triboluminescence, was studied in detail and some were tested out in a home laboratory. The study on control of light encompassed reflection, diffraction, dispersion, deflection, and went into mechanical shutters, vibrating mirrors, the Kerr cell, birefringence, and polarization. A small telescope mirror was hand-ground and Schlieren optics studied. This study was mainly library research and covered several years.

About that time Marks of the Marks Polarizing Company and Land of Polaroid were publishing their work on producing polarizers from quinine iodosulfate. Marks was growing large thin crystal sheets and Land was stretch-orienting small needle-like particles.

The key to the development of these polarizers was to recognize that they function by absorbing light vibrating in one direction and transmitting light vibrating in the perpendicular direction. They are dichroic.

In looking up articles on dichroism the work of Dr. Hans Zocher came up. He was working with the alignment of dichroic materials by rubbing them. He was also working with dyes that had a liquid crystal state whereby they were oriented by the rubbed pattern of the substrate upon which they were in contact while in the lyotropic liquid crystal state. He also had found that he could obtain light polarizers by oriented particles of some of these same dyes and obtained a U.S. patent.

The phenomena that was intriguing was the epitaxial character of orienting on a rubbed surface. This appeared to be a unique characteristic of liquid crystals. Dr. Zocher had published this finding as part of a paper but Dr. P. Chatelain, at a somewhat later date, had published a paper on the same phenomena, without knowledge of Dr. Zocher's work. Dr. Chatelain gets the credit for being the original discoverer of the rubbed-surface-orienting phenomena, for as Dr. Zocher graciously said, "Who discovered America? Was it the early Indians, the Vikings, or Columbus? The credit for the discovery goes to he who makes the world know about it, not necessarily to he who was first."

By a simple laboratory experiment using a different technique of fast-drying and one of Dr. Zocher's liquid crystal dyes, it was found that the orientation due to the rubbed glass substrate could be maintained as a solid film and not go into the oriented particle state.

Here was something apparently new and potentially useful. It looked like it had the possibility of being developed as a process that had unique characteristics, one that could make commercially useful products. Furthermore, it was much simpler than the original idea. The only other commercial product using liquid crystal phenomena was soap.

I made the decision to go ahead with the development of a light polarizer using the liquid crystal state. The investigation then shifted from mainly literature to mainly laboratory research. It was a month before repeated tries reproduced the original experimental results. It then took 10 years of odd hour experimentation before a product was to be commercialized.

Dr. Kettering has said that frustrating experimental results are but beacon lights trying to tell you the truth. Nature never lies to you. That is true; but nature also holds onto its secrets very tightly. Faith in the inherent correctness of what you are doing eventually pays off and when you get the real answers they have beautiful clarity.

The commercial product had to have light fastness, go from black to white, etc. Patents were obtained as these features were developed. The first application I developed was a way to eliminate glare from incandescent light bulbs [see Figure 16.1]. At that time these were mainly of the spherical type which had a curved surface and I thought that this liquid application to a curved surface had advantages over the flat sheets of the competition. I was unsuccessful in finding a buyer for the patents and process. I decided to try to proceed with manufacture and sale of these lamps. However, because dichroic polarizers operate by absorption more than half the light, their use on lamps was not a commercial success. Later, when three-dimensional movies came along millions of viewers were made using the same process, but that is another story.

A STUDENT'S FIRST INVENTION

Philip Rilinger wrote the following story of his experience as an inventor in response to an assignment while taking my course in product design during the autumn quarter of 1979. Phil was a senior then. The events related in the story started when he was a prejunior (third year of UC's five-year co-op program).

While watching the movie "Star Wars" at the local theater, one scene particularly affected me. White-clad aliens were shooting

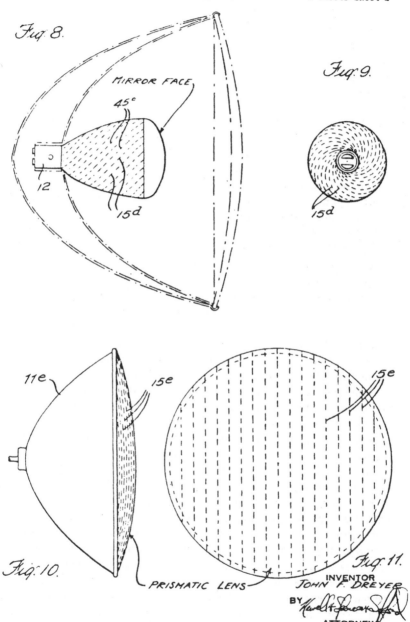

Fig. 8.

MIRROR FACE

45°

12

15d

Fig. 9.

15d

11e

15e

15e

Fig. 10.

PRISMATIC LENS

Fig. 11.

INVENTOR
JOHN F. DREYER

BY

ATTORNEYS

Figure 16.1   John F. Dreyer's patent (U.S. Patent Office).

127

each other with a strange-sounding gun that shot a beam of light. The thought occurred to me, wouldn't it be neat for a little kid to have such a gun that produced the same sound. The problem with this idea was the fact that such a sound was being created by a very expensive music synthesizer. How could one come up with the same sound inexpensively?

My first step was to observe what the sound looked like. This I did by hooking an oscilliscope up to my stereo and playing that particular part of the sound track. After trying many times to isolate this sound from others, I concluded that the sound looked like a damped sinusoid.

From a previous course in college, a simple underdamped RLC circuit seemed as if it would be the perfect thing to realize this shape. Doing preliminary calculations with some standard values for R, L and C, I soon found it very difficult to come up with a 16kHz wave that died out in a relatively short time. It also seemed that using an inductor would make things awkward and expensive. By this time exams were approaching and I decided to abandon the whole project.

Working at a co-op job a couple of weeks later, I walked by a worker who was tearing a sheet of thick plastic. The sound that it made was just the thing I'd been looking for. Being an electrical engineering student, the thought of doing it mechanically never entered my mind. Running through some different ideas, nothing seemed to work and once again I forgot about it.

Some time later in an electronics class, I was introduced to a circuit called the Wein Bridge oscillator. The frequency could be easily adjusted using Rs and Cs and best of all no inductors were needed. Interest was once again rekindled and the idea occurred to me to modulate the sine wave output with a transistor and capacitor arrangement. Finally I had the sound and what I thought was a fairly inexpensive circuit to build [see Figure 16.2].

By this time the movie was a big hit and a local toy manufacturer had obtained the rights to manufacture all toys related to the motion picture. I made contact with the manufacturer and asked if they would be interested in an idea related to the movie. To my dismay they set up an appointment for me to talk with them.

Legal ramifications never entered my mind and so it surprised me when I was presented with something called a letter of disclosure to sign, the letter supposedly protecting the company and individual. They looked at the model and liked it, but rejected it on the

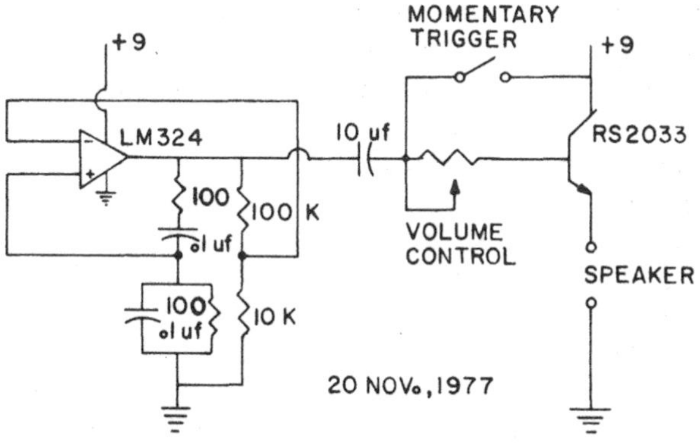

Figure 16.2   A student's invention: sounds of "Star Wars." (By permission of Philip Rilinger.)

basis that their company was also working on something similar. They wanted to keep a copy of the circuit diagram as spelled out in the letter of disclosure.

This experience taught me how important a patent is and how difficult it is to get a toy accepted.

Looking back through the steps I used in this process it surprises me as to how close the things I did related to the principles outlined in the pamphlet "Alternative Designs and Inventions."*

A LIFESAVING INVENTION

Ben Horvay came to this country from Hungary in 1940 [36]. Three years later he received the degree of Bachelor of Science in Electrical Engineering from the University of Cincinnati after an accelerated program of study and cooperative work experience. Most of his career since then has been with the General Electric Company. In the 33 years of service to that organization he has been the inventor of 33 patented improvements.

Besides working in product design and technical leadership, Ben is also a teacher. He has taught thermodynamics at the University of Louisville since 1962 and was named Adjunct Associate Professor of Mechani-

---

*A student handout used in the product design course which, with the help of students and others, became this book.

cal Engineering in 1971. Within the GE company he has taught courses in basic and advanced refrigeration to hundreds of refrigerator and air conditioner engineers and served as class supervisor in that company's well-known Creative Engineering Program in the 1940s.

In 1979 he received the Charles P. Steinmetz Award in the Consumer Products and Services Sector of General Electric. He has also received the Higgins Award of the Pressed Metal Institute, the Appliance Engineers Society Award, and the Distinguished Alumni Award of the University of Cincinnati. He is a Fellow of the American Society of Heating and Refrigeration Engineers. Ben Horvay's professional career has enabled him to make significant contributions to safety, lower costs, higher quality, and greater standardization in the products he has touched. His teaching has enabled him to share his rich technical insight with those who want to learn. Here is one example of his contributions.

I truly believe that creativity is a God-given talent that each human being is blessed with at birth, some to a larger extent than others. Nevertheless, everyone of us can nurture and develop this talent by a systematic approach to problem solving. The true creative genius, be it a composer, a painter, or an engineer, follows such a systematic approach by native instinct with fanatical perserverence and dedication. We lesser souls must force ourselves to do the same, until it becomes second nature. Such a systematic approach can be *taught* and can be *learned*. I consider myself a living example of how one's limited native creative talent can be harnessed, disciplined, and optimized to produce useful and simple solutions to seemingly complex problems.

When I joined General Electric's Household Refrigerator Department in 1950, refrigerator door latches were mechanical, over-center devices in which a spring was compressed by the actuation of the door handle when opening the door, and tripped or released by the strike when closing the door.

There were several problems with these mechanical latches: First, the latch could be tripped with the door in the open position with the result that it would not latch on to the strike, but bounce back open. The customer would have to cock the latch by operating the door handle in the open position, something that did not come naturally to many persons and often resulted in service calls. Mechanical latches also required individual adjustment of the strike, both in the horizontal and in the vertical plane to assure proper positioning with respect to the latch. Noise was objectionable. There was also an ever increasing concern with child entrapment. Newspaper stories describing the suffocation of children playing in abandoned refrigerators were pointing an accusing finger at the refrigeration industry. Public concern with child entrapment resulted in considerable research by the National Bureau of Standards into the behavior of small children when entrapped and into the forces that they could exert from the inside of a refrigerator to open the door. When Public Law 930 finally issued in 1956, this force was specified as not to exceed 15 pounds when exerted on the latch side of the door. Mechanical latches of the day would not budge even at forces tenfold that amount.

The need for such a great latching force was due to the bimetallic nature of the refrigerator door assembly. The coefficient of thermal expansion or contraction of the plastic inner door is considerable. The plastic inner door is cooled down from room ambient to 38 °F, or to 0 °F depending on fresh food or freezer application, causing it to contract while the steel outer door expands, although to a lesser extent, in a warm kitchen. This results in the bowing of the entire assembly by as much as 1/4 in. at the center of the door if unrestrained by the latch, and a loss of the door seal occurs.

The first approach to overcome these problems was the magnetic gasket, invented by Mr. Alfred Janos, my boss and mentor in the early 1950s at General Electric [Figure 16.3]. Individual ALNICO 5 magnets were located inside a special pocket of the door gasket with like poles adjacent as shown. The steel outer case of the refrigerator cabinet acted as the magnetic keeper with the door in the closed position. The magnetic gasket solved a lot of problems and introduced some new ones. It eliminated the need for strike adjustment, proved to be child-safe in the spirit of Public Law 930, and could not be tripped accidentally like the mechanical latch. Most significantly, it responded well to door bowing, by the gasket stretching instead of compressing, assuring good seal under a variety of environmental conditions.

There were also new problems. There was a poor door seal at the corners because of the stiffness of the miters (and this is where the magnetic force was the weakest), the gasket material failed early due to repeated stretching, and last but not least, iron particles from the air were attracted by the magnets. These particles clung to the surface of the gasket causing undue wear and eventual corrosion of the front flange of the cabinet. My task was to modify the design to overcome these problems.

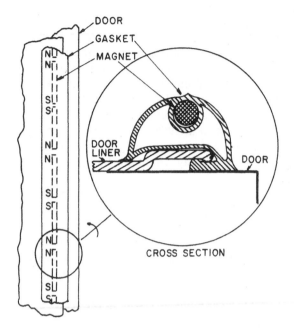

Figure 16.3 Early design of the magnetic gasket. (By permission of J. Ben Horvay.)

I followed my usual 5-step procedure:
Step 1 Definition of the problem
Step 2 Search for methods
Step 3 Illumination
Step 4 Evaluation of alternates
Step 5 Decision

The most critical step is 3, illumination. This is where invention occurs, probably in the subconscious (or preconscious) mind. To prepare the ground for it, I force myself to generate at least seven alternates in step 2, while I search for a method of solution. There are tricks I use to aid in generating alternates, such as substitution of materials, rearranging the parts, combining functions, multiplying or magnifying certain elements, or subtracting, minifying, eliminating components. This last exercise led me to the single magnet concept and the idea of having one large magnet instead of a lot of small magnets. The original design was a monstrosity [Figure 16.4, U.S. Patent 2,662,787]. The magnet (part 9) was so powerful that a

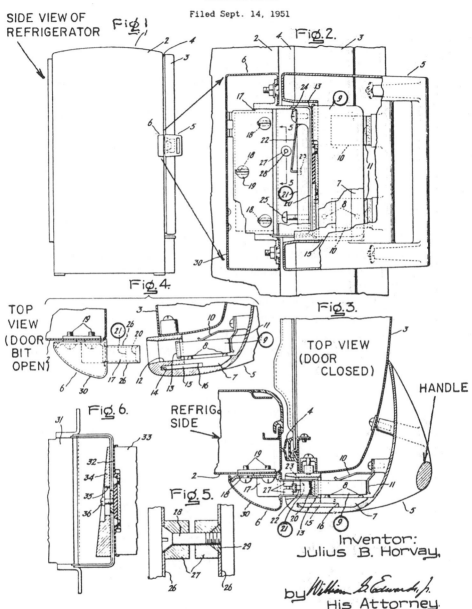

SIDE VIEW OF REFRIGERATOR   Fig.1

Fig.2.

Fig.4.

TOP VIEW (DOOR BIT OPEN)

Fig.3.

TOP VIEW (DOOR CLOSED)

HANDLE

Fig.6.

REFRIG. SIDE

Fig.5.

Inventor:
Julius B. Horvay,

by William G. Edwards, Jr.
His Attorney.

Figure 16.4   J. Ben Horvay's magnetic latch. (By permisssion of J. Ben Horvay.)

134

shield was provided to keep customers' wristwatches from being magnetized. An adjustable keeper (part 21) was incorporated to provide adjustable magnetic pull.

By the time the magnetic latch went into production in 1955, it was greatly simplified and improved [Figure 16.5]. The magnet was of simple horse shoe shape and the customer saw only the escutcheons and eyelets protruding thru the plastic inner door [Figure 16.6]. One major contributing factor to this simplified design was the development of significantly softer door gaskets. These required much less compressive force and were more tolerant to door bowing.

The magnetic latch provided all the benefits of the original magnetic gasket (individual magnets inside the door gasket) and overcame its problems. Using a compression gasket it provided excellent corner seal, it did not tear, and it did not collect steel particles except at the magnet tips. (The keeper was provided with a stainless steel cover to protect the cabinet from being damaged by steel particles there.)

When the magnetic latch was first applied to a food freezer rather than to a refrigerator, it was a disaster. After an hour or so of operation the door suddenly popped open—due to door bowing, which is much more pronounced with a freezer because of lower in-

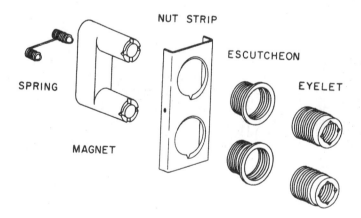

Figure 16.5 Simplified, one-magnet latch. (By permission of J. Ben Horvay.)

Figure 16.6 Refrigerator with simplified, one-magnet latch. (By permission of J. Ben Horvay.)

side temperatures. This forced me to develop a floating magnetic keeper [U.S. Patent 2,812,965] that moved out with the magnet as the door bowed.

In recent years magnetic gaskets reappeared on the scene using continuous ferrite magnets [Figure 16.7]. These gaskets are much superior to the original version with individual magnets, and have been slowly replacing magnetic latches.

Magnetic door closures, whether gasket or latch, have overcome one of the most tragic accidents imaginable. While with mechanical latches 15 to 20 accidents of child entrapment in abandoned refrigerators were reported annually, I am aware of no such incident that involved a refrigerator or freezer using a magnetic door closure.

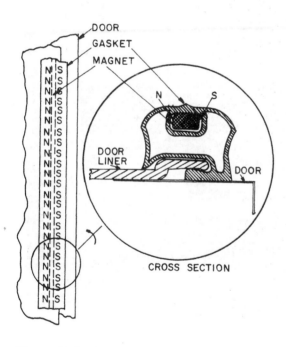

Figure 16.7 Later design with soft gasket. (By permission of J. Ben Horvay.)

17
Closure

A PARTING WORD

This book presents information which I hope will motivate you to become interested in invention. For many readers it may have stripped away the mystery of inventing to put it within grasp. One of its main messages is that a successful invention is often a conceptually small but commercially important improvement which fills a widespread need. The first step is to recognize such a need.

The book also attempts to lead you to understand creativity and what you must do to become a more creative engineer. Nothing comes from a vacuum. You should work diligently to follow the advice in Chapter 6. This is probably the most important chapter in the book.

Finally, plan your campaign to find a buyer with care and do not become trapped in the pessimism of defeat. Not every bright idea or even every patent that holds your name will be financially rewarding. Unless you intend to live only by inventing you need but few successes. Just as important, however, are the intangible benefits. I have never met an inventor who was not happy about being one.

EXERCISES

Usually books of this series do not have exercises for the reader to pursue. However, I believe it will be helpful in this case. The first group is meant to motivate you to review what you have read. The second group gives invention suggestions. These are meant to be realistic suggestions of needs which could be rewarding to the inventor.

TO LEARN IT BETTER

1. Other methods of stimulating invention may be found in the literature. Make a library search on the subject of creativity and invention and write a synopsis of at least one method not included here.
2. What change of strategy did Harold Black make in his attempt to reduce amplifier noise? Was it a key step in his successful development of the feedback amplifier?
3. Review the examples of invention given in the book and make a list of the lesson or lessons to be learned from each example.
4. What fact related in the story of Kestin solving the problem of the right triangle other than the occurrence of bisociation substantiates the scenario of creativity described in this book?
5. In the case history of Phil Rilinger's invention, list the ways in which his actions deviated from what you would do after having read this book.
6. Compare John Dreyer's development of the light polarizer to the landmark inventions given in Chapter 3. Which of those stories is most like John's?
7. Find three examples of elegant inventions in your home and study them to determine why they are superior. An example is the lid design of the pretzel and potato chip containers made by Tupperware. These are airtight to keep the contents fresh. The container and lid are made of thermoplastic and the lid is so designed that, when one pushes down on the middle, the outer rim contracts. This allows it to fit the upper edge of the container and, when the center force is removed, the rim of the lid presses against the inner surface of the top edge of the container for the necessary tight fit. This is an elegant invention because it is easy to remove the force-fitting lid, easy to put the lid in place by pressing it correctly, and it costs no more than other plastic lids, even those that are not airtight.
8. Make a list of four needs that you recognize in your home or business situation which could lead to invention.
9. Make a list of the characteristics of creative people given in Chapter 4 and compare it to your characteristics as you perceive them.
10. Of the 15 suggestions given in Chapter 6 on how to improve your creative ability, list those which seem most important for you to pursue.
11. Identify the most important step in each method to stimulate invention given in Chapters 5 and 8.
12. Make a first try at developing a method to stimulate invention which you feel may be beneficial to *you*.

INVENTION SUGGESTIONS

1. One of the survey respondents suggested that a residential door bell be designed for total automatic machinery manufacturing. The history of the electric door bell is that in the 1920s and early 1930s the device was indeed a bell with the hammer actuated by an electromagnet which broke the circuit as the strike occurred and allowed a spring to return the hammer to the open position for the next strike. Then it became popular to use chimes, in which plungers strike bars which are resonant at different frequencies. I have found these to be unreliable. Visitors call and remain unannounced because the plunger friction has increased. Lighter fluid applied to the plunger corrects the problem but the question of how long the door chimes did not work before the problem was discovered is annoying. Thus, you could meet a need by developing a device which would be reliable without maintenance and inexpensive because it can be made by automatic machinery.

2. Invent a burglar alarm system to be installed in older homes. This is a particularly difficult problem because the installation of electric wires to doors and windows is expensive.

3. Drive-through car washes are becoming expensive and the care with which the job is done seems to be decreasing. Since many families have two or more cars, it seems reasonable that a home washer costing perhaps as much as $100 would enjoy considerable sales. Attempts have been made to use water from high-speed nozzles to get an inexpensive, easy way to wash a car. This technique does not remove all dirt. Some form of mechanical rubbing or brushing seems to be necessary. An intriguing idea is to use water pressure to provide the brush-moving action because generous use of water helps remove the abrasive dirt. Also, ideally the device would allow the user to avoid getting his clothes wet.

4. Toasted bread varies considerably from piece to piece in degree of carbonization. The control is operated by heat. People typically comment that the toast is too dark or too light, i.e., they evaluate it by color. Develop a method by which the user could choose the color.

5. Measurement of various physical constants will allow you to test your creative ability. For example, show by sketch suitable equipment to conveniently measure (1) static friction, (2) specific gravity, (3) specific heat, (4) surface hardness, and (5) other constants.

6. In squeezing material from a tube the material continues to exit after your hand pressure is stopped. Some materials, e.g., epoxy,

are expensive enough that a manufacturer who could advertise that his material is in a retractable tube would have a commercial advantage. Is there an inexpensive way to improve a tube so that it pulls material back into itself when released?

7. Residential streets off of main thoroughfares usually have stop lights at the entrance that are tripped by a car as it arrives at the main street for ingress. I have noticed that cars coming the other way, i.e., entering the residential street from the thoroughfare, often take such a wide sweep that they trip the light. As a result traffic on the busy street is stopped when no car is waiting to enter it. Invent a traffic light control that detects the direction in which a car is passing over it.

8. There is much talk about increasing productivity. This is best done for the factory worker by providing him with better production machinery. But "white-collar" workers, viz., executives, professional people, and office personnel, need to increase productivity as well. The first task is for them to find out how they spend their time. It may be quite different from how they believe they spend it.

One way to make a scientific determination of activities is to record at random times what they are doing. Consider a device with 10 buttons to record 10 or fewer activities (one can be labeled miscellaneous). This device is made to signal by light or audible sound, or both, when the activity at that moment should be recorded. The recording requires only pushing one of the 10 buttons. The signal to record must be truly random. In fact, the user should be able to judge that the test has accumulated enough data when the signals which occurred during lunch hours, and were cleared upon return to the workplace by pressing the button marked "lunch," are one-ninth of the total recordings for work days of nine hours which includes one-hour lunch periods.

9. Tire wear depends upon a correct parallel alignment of the front wheels. It would be reassuring to have a method of checking the alignment on a do-it-yourself basis. This is especially true because, unlike most car maintenance, even after service there is no way to tell whether the adjustment was made. No matter how often you ask for the alignment to be checked you car always seems to need alignment. Invent an inexpensive throw-away device which can be used to determine wheel alignment at home or a reasonably priced device which would last indefinitely.

10. Clothes even if dry-cleaned retain the characteristic odor of having

been worn. Dry cleaning does not freshen clothes as washing with detergent and water does. Can you improve the process to give a newly dry-cleaned suit the freshness of one just off the clothing store rack?

11. Buildings, stone walls, and sidewalks are often defaced these days by persons using a spray can of paint to write messages for all to see. Develop a technique for removing this damage.

12. When all is said and done you are most likely to make a worthwhile invention by recognizing a need based upon your special situation. Consider your home, your job, and your hobbies. Make a list of all the needs you can identify. Select the one most attractive to you and try for the elegant way to fill it. Keep the list current. Add to it when you recognize a new need. Strike out those which after a time seem trivial. Don't just sit there; invent something!

References

1. Commissioner of Patents and Trademarks Annual Report Fiscal 1978. Government Printing Office, Washington, D.C. Interpretation of data contained therein.
2. F. B. Dent, Technology Assessment and Forecast: Early Warning Report, Office of Technology Assessment and Forecast, Washington, D.C. December, 1973.
3. D. V. DeSimone, (Ed.), *Education for Innovation,* Pergamon, New York, 1968, pp. 23–45.
4. A self-locking nut that fits anywhere *Business Week,* November 22, 1976, p. 82J.
5. Official Gazette of the United States Patent Office, issued weekly. Superintendent of Documents, Government Printing Office, Washington, D. C. 20402. Also available in many university and public libraries.
6. 'Little guy' battles Sears, wins $1 million in suit. *Cincinnati Enquirer,* October 12, 1978, p. A6.
7. C. F. Dalziel, Electric shock hazard, *IEEE Spectrum,* 9(2): 41–50, February 1972.
8. C. F. Dalziel, Transistorized ground-fault interrupter reduces shock hazard, *IEEE Spectrum,* 7(1): 55–62, January 1970.
9. W. G. Konold et al., *What Every Engineer Should Know About Patents,* Marcel Dekker, New York, 1979, pp. 11, 48, 58–64.
10. Geoffrey Smith, It's better than chasing girls, *Forbes,* 124(12), December 10, 1979, pp. 85–89.

11. R. Donaldson (Ed.), *Proceedings of Bicentenary of the James Watt Patent,* University of Glasgow, 1969, p. 6.

12. Xerox, Interim Report to Stockholders, September 30, 1978.

13. H. S. Black, Inventing the negative feedback amplifier, *IEEE Spectrum,* 14(12): 54–60, December 1977.

14. R. M. Page, *The Origin of Radar,* Anchor Books, Doubleday, New York, 1962, pp. 122–123.

15. P. E. Vernon (Ed.), *Creativity,* Penguin, Baltimore, 1970, pp. 53–83, 107–151, 289–311.

16. A. Koestler, *The Act of Creation,* Danube Edition, Macmillan, London, 1969, pp. 101–107.

17. J. Kestin, Creativity in teaching and learning, *American Scientist,* 58: 250–257, May–June 1970.

18. *Encyclopaedia Britannica,* Vol. 10, 1971, p. 1052.

19. J. Greenberg, Einstein: The gourmet of creativity, *Science News,* 115, March 31, 1979, pp. 216–217.

20. L. S. Kubie, Blocks to creativity, *International Science and Technology,* 42: 69–78, June 1965.

21. R. Hyman and B. Anderson, Solving problems, *International Science and Technology,* 45: 36–41, September 1965.

22. F. Barron, *Creative Person and Creative Process,* Holt, Rinehart and Winston, New York, 1969, p. 102.

23. H. C. Lehman, *Age and Achievement,* Princeton University Press, 1953.

24. D. B. Bromley, Experimental tests of age and creativity, *Journal of Gerontology,* 11: 74, 1956.

25. E. J. Tangeman, Creativity, the facts behind the fad. *Product Engineering,* August 24, 1959, pp. 20–23.

26. R. P. Alley and C. F. Hix, Jr., *Physical Laws and Effects,* John Wiley and Sons, New York, 1958.

27. W. E. Kock, *The Creative Engineer,* Plenum, New York, 1979, pp. 166–170.

28. J. Rabinow, 1980 Scientist of the year lecture *Industrial Research and Development,* 22(11): 108–112, November 1980.

29. R. L. Bailey, *Disciplined Creativity for Engineers,* Ann Arbor Science Publishers, Ann Arbor, Michigan, 1978, pp. 422–435, gives a list of 135 methods.

30. A. F. Osborn, *Applied Imagination,* 3rd ed., Charles Scribner's Sons, New York, 1963.

31. W. J. J. Gordon, *Synectics, The Development of the Creative Capacity,* Harper and Row, New York, 1961.

144

32. Business probes The creative spark, *Dun's Review,* 115(1): 32–38, January 1980.

33. R. E. Manelis, Enriching creative talents, *Automation,* 17 (II 7): 34–41, July 1970.

34. W. H. Middendorf and G. T. Brown, Orderly creative inventing, *Electrical Engineering,* 76(10): 866–869, October 1957.

35. F. Zwicky, *Discovery, Invention, Research Through Morphological Approach,* Macmillan, New York, 1966, pp. 25, 250–260.

36. J. B. Horvay, G. E. inventors: What Makes Them Tick? General Electric Company internal publication.

37. G. Kivenson, *The Art and Science of Inventing,* Litton, New York, 1977, pp. 19–20.

38. *Encyclopaedia Britannica,* Vol. 10, 1971, p. 575.

39. F. S. Wickware, The amazing Dr. Sperti, *Collier's,* July 29, 1950, pp. 25, 72–73.

40. G. S. Sperti, private correspondence.

41. H. A. Liebhafsky, *Silicones Under the Monogram, A story of Industrial Research,* John Wiley and Sons, New York, 1978, pp. 317–322.

42. For example: How Creative Are You? Princeton Creative Research, Inc., 10 Nassau Street, P.O. Box 122, Princeton, N.J. 08540.

43. E. Raudsepp, Testing for creativity, *Machine Design,* 37(15), June 24, 1965, pp. 122–128.

44. E. Raudsepp, Games that stimulate creativity, *Machine Design,* 49(17), July 21, 1977, pp. 88–94.

45. *Science and Invention Encyclopedia,* Stuttman, International Ed., 1977, Vol. 7, pp. 834–835.

46. *Science and Invention Encyclopedia,* Stuttman, International Ed., 1977, Vol. 19, p. 2541.

47. M. F. Wolff, Inventing at breakfast, *IEEE Spectrum,* 12(5): pp. 44–49, May 1975.

48. W. H. Middendorf, Standards—the evidence of concern, *IEEE Spectrum,* 8(8): pp. 70–73, August 1971.

Index

T - #0579 - 101024 - C0 - 229/152/9 - PB - 9780824774974 - Gloss Lamination